Civil engineering in context

Alan Muir Wood

🍸 Thomas Telford

Published by Thomas Telford Publishing, Thomas Telford Ltd, 1 Heron Quay, London E14 4JD.
www.thomastelford.com

Distributors for Thomas Telford books are
USA: ASCE Press, 1801 Alexander Bell Drive, Reston, VA 20191-4400
Japan: Maruzen Co. Ltd, Book Department, 3–10 Nihonbashi 2-chome, Chuo-ku, Tokyo 103
Australia: DA Books and Journals, 648 Whitehorse Road, Mitcham 3132, Victoria

First published 2004

A catalogue record for this book is available from the British Library

ISBN: 0 7277 3257 9
© Alan Muir Wood 2004

All rights, including translation, reserved. Except as permitted by the Copyright, Designs and Patents Act 1988, no part of this publication may be reproduced, stored in a retrieval system or transmitted in any form or by any means, electronic, mechanical, photocopying or otherwise, without the prior written permission of the Publishing Director, Thomas Telford Publishing, Thomas Telford Ltd, 1 Heron Quay, London E14 4JD.

This book is published on the understanding that the author is solely responsible for the statements made and opinions expressed in it and that its publication does not necessarily imply that such statements and/or opinions are or reflect the views or opinions of the publishers. While every effort has been made to ensure that the statements made and the opinions expressed in this publication provide a safe and accurate guide, no liability or responsibility can be accepted in this respect by the author or publishers.

Typeset by Academic + Technical, Bristol
Printed and bound in Great Britain by MPG Books, Bodmin, Cornwall

To Winifred.

Inspiration and support
over many, many years

Contents

Abbreviations and acronyms   viii

Introduction   xi

Chapter 1   An historical perspective   1
1.1   Introduction, 1
1.1.1   Hagiographic, 1
1.1.2   Utilitarian, 2
1.1.3   Educational, 3
1.2   Uses of history, 4
1.3   The Greeks and the Romans, 6
1.4   The Industrial Revolution, 10
1.5   The 'Golden Age' and decline, 12

Chapter 2   Recent historical perspective   22
2.1   Dominant features, 22
2.2   The so-called 'traditional method' of project procurement and management, 24
2.3   What went wrong with the ICE Conditions?, 27
2.4   Recent reports affecting construction, 29
2.4.1   Banwell Report, 29
2.4.2   Latham Report, 29
2.4.3   Egan Report, 34
2.5   The New Engineering Contract, 35
2.6   The FIDIC Suite of *Contract Documents* (1999), 37
2.6.1   Provisions for continuity, 38
2.6.2   Relationships between the Parties, 38

2.6.3 Value engineering, 39
2.6.4 Provisions for uncertainty, 40
2.6.5 Conclusions on red FIDIC, 41
2.7 Highways Agency Early Contractor Involvement Scheme, 42

Chapter 3 Engineering and the institutions   43
3.1 The wider scene of engineering, 43
3.2 Unity of the engineering profession, 49
3.3 Evolution of the Institution of Civil Engineers, 51
3.4 ICE Policy on reports and publications, 54
3.5 Qualifications for membership, 57
3.6 An historical perspective for the profession, 58

Chapter 4 Engineering, management and the law, 60
4.1 Introduction, 60
4.2 Principles of good and bad practice, 61
4.3 Decline of the ICE Conditions, 62
4.4 Application of principles of good practice, 63
4.5 Getting the best out of Contract Conditions, 66
4.6 The engineer and the law: pre-project, 73
4.7 The engineer and the law: the project and beyond, 75
4.8 The expert witness, 77

Chapter 5 Systems and design   83
5.1 Systems engineering, 83
5.2 Definition of the project, 92
5.3 Observation and anticipation, 95
5.4 The Observational Method, 99
5.5 The use of models, 103
5.6 Design for the future, 109
5.7 The role of the Proof Engineer, 110

Chapter 6 Learning from experience   113
6.1 First impressions, 113
6.2 University – the foundation for curiosity, 115
6.3 The Navy at war, 117
6.4 A start at civil engineering, 121
6.5 Folkestone Warren, 124
6.6 Another touch of research, 126
6.7 In consulting practice, 127
6.7.1 Tunnelling, 127

6.7.2	Coasts, estuaries and landslides, 134	
6.7.3	Silos and miscellaneous structures, 139	

Chapter 7 The engineer for the twenty-first century — 145
7.1   The historical legacy, 145
7.2   Repositioning of the civil engineer, 150
7.3   Satisfaction as a career, 152
7.4   Engineer's code of conduct, 155
7.5   Implications for education and training, 158
7.6   The last days of pragmatism, 163

Chapter 8 Channel Tunnel — 166
8.1   The revival of interest from 1958, 166
8.2   The Tunnel becomes a reality, 172
8.3   Procedural defects of decision-making, 175
8.4   Comparison with the Øresund project, 177
8.5   Early history of the Channel Tunnel, 178
8.6   Random technical reflections, 179

Chapter 9 Ethics and politics — 182
9.1   Introduction, 182
9.2   Engineering ethics, 184
9.3   Engineering ethics translated into politics, 190
9.4   Personal excursions into the political scene, 194
9.4.1 ACORD, 195
9.4.2 ACARD, 196
9.4.3 The launch of the International Tunnelling Association, 197
9.4.4 Research Councils, 199
9.5   The engineer, the politician and the public interest, 201
9.5.1 Private and public, 204
9.5.2 Railways, 206
9.5.3 Energy, 208
9.6   The environment and sustainability, 210
9.7   Broader reflections, 212

Chapter 10 CODA — 219

References — 221

Index — 227

Abbreviations and acronyms

		First encounter
ACARD	Advisory Council on Applied Research and Development	9.4.2
ACORD	Advisory Committee on Research and Development	9.4.1
AITES	Association Internationale des Travaux en Souterrain	9.4.3
ALARP	As low as reasonably practical	8.3
ASCE	American Society of Civil Engineers	3.4
BAA	British Airports Authority	4.5
BCSA	British Constructional Steelwork Association	6.7.3
BOT	Build–operate–transfer	2.1
bpm	best practicable means	9.6
BRS	Building Research Station	7.1
BS	British Standard	2.4.2
CASE	Court Appointed Scientific Experts	4.8
CCLM	Centre for Construction Law and Management	4.1
CDM	Construction Design Management	2.4.2
CEGB	Central Electricity Generating Board	6.7.2
CEI	Council of Engineering Institutions	3.1
CEng	Chartered Engineer	3.2
CERA	Civil Engineering Research Association	7.1
CETu	Centre d'Etudes des Tunnels	9.4.3
CFD	Computational Fluid Dynamics	5.5
CIRIA	Construction Industry Research and Information Association	2.4.2
CPR	Chartered Professional Review	3.5

Abbreviations and acronyms

CPR	Civil Procedure Rules	4.8
CSR	Corporate Social Responsibility	9.1
CTSG	Channel Tunnel Study Group	8.1
DAB	Dispute Adjudication (or Advisory) Board	2.6.2 (8.4)
DBFI	Design, Build, Finance and Transfer	9.5.2
DSIR	Department of Scientific and Industrial Research	7.1
DTI	Department of Trade and Industry	3.1
EC	(and EC^{UK}) Engineering Council	3.1
ECI	Early Contractor Involvement	2.7
EIA	Environmental Impact Assessment	9.2
EPC	Engineering, procurement, construction	2.6
ERA	Engine Room Artificer	6.3
ESRC	Economic and Social Research Council	9.4.4
ET	Eurotunnel	8.2
ETB	Engineering and Technology Board	3.1
FIDIC	Fédération Internationale des Ingénieurs Conseils	2.3
GEC	General Electric Company	9.7
GBR	Geotechnical Baseline Report	4.5
GDP	Gross Domestic Product	9.5.1
GPS	Geographical Positioning Systems	6.7.1
HA	Highways Agency	2.7
HND	Higher National Diploma	3.1
HRS	Hydraulics Research Station	7.1
HSE	Health and Safety Executive	6.1
ICC	International Chamber of Commerce	4.7
ICE	Institution of Civil Engineers	1.1
IEng	Incorporated Engineer	3.2
IPR	Incorporated Professional Review	3.5
IStrucE	Institution of Structural Engineers	7.5
IT	Information Technology	3.5
ITA	International Tunnelling Association	9.4.3
ITDG	Intermediate Technology Development Group	7.2
JBM	Joint Board of Moderators	6.2
MICE	Member of Institution of Civil Engineers	3.5
MoT	Ministry of Transport	7.1
MW	Megawatt	5.5
NCE	New Civil Engineer	2.5
NEC	New Engineering Contract	2.5
NIREX	Nuclear Industry Radioactive Waste Executive	6.7.1
NPL	National Physical Laboratory	7.1

OECD	Organisation for Economic Co-operation and Development	9.4.3
OM	Observational Method	5.4
PBE	Professionals for the Built Environment	5.1
PFI	Private Finance Initiative	4.5
PIANC	Permanent International Association of Navigation Congresses	7.1
PIARC	Permanent International Association of Road Congresses	7.1
QA	Quality Assurance	2.4.2
RedR	Register of Engineers for Disaster Relief	7.2
RIBA	Royal Institute of British Architects	3.1
RRL	Road Research Laboratory	7.1
SARTOR	Routes to Qualification of Professional and Technical Engineers	3.1
SERC	Science and Engineering Research Council	9.4.4
SNCF	Société Nationale des Chemins de Fer	8.1
SRA	Strategic Rail Authority	9.5.2
TA	Territorial Army	8.1
TBM	Tunnel Boring Machine	6.7.1
TML	Trans-Manche Link	8.2
TRRL	Transport and Road Research Laboratory	7.1
TRL	Transport Research Laboratory	7.1
UNESCO	United Nations Educational and Scientific Committee	3.6

Introduction

Writing a book entails much perusal of sources and occupation of time that might be enjoyably spent in other pursuits. The author must therefore be able to justify, at least to himself – the alternative of herself, he or she, his or her, is to be understood throughout the text – valid reasons for so doing. In the absence of this degree of self-assurance, readers are unlikely to be persuaded to penetrate far beyond the first paragraph of the Introduction.

I have enjoyed, and for the most part the verb is apt, a long career in engineering. I have seen many changes, some for better, some for worse, some within the profession, many without. As I recount in this book, from a time well before preparation to become an engineer, I began to suspect that technology in Britain had stalled. To understand the reasons, and to hope to make a contribution to overcome them, provided triple spurs of curiosity, challenge and ambition. These must surely be the incentives for any engineer.

I do not doubt that we possess in our profession many strengths, actual and potential, the latter remaining to be identified in order to be realised. Our weaknesses could be readily corrected. Technical engineering has raced ahead of practical management, which engineers have allowed to fall excessively into the hands of others who do not understand the criteria for success. Social and environmental factors have assumed, and will continue to assume, increasingly important influences on all the engineers' decisions. Throughout engineering, there is a general lack of purpose to provide the resources necessary to carry research into effective application through innovation, which is what technology is all about.

The engineer is essentially a synthesist, joining together different elements derived from different sources into a system. The organisation of the processes that constitute the system requires the highest capability of engineering. Innovation is universal, often achieved without intent; this will occur even where no new idea is introduced, since the blending of the familiar, at least for a civil engineer, will be different from one project to another. It is often the correct identification of the similarities and differences between applications that requires skilled discernment, and is vital for success.

While preparing mentally to write this book, I was struck, as many engineers have been, by the theme of 'Trust' of the 2002 BBC Reith Lectures by Onora O'Neill. My own emphasis throughout this book is on the benefit that derives from the proper use of, and respect for, professionals. Trust is in fact the counterpart of ethical behaviour in general and engineering ethics in particular. Thus, the professionals can only be used to the full where they merit this trust. In professional terms this goes far beyond mere probity, although this is clearly a vital feature. The essence is that of one thoroughly competent for the job in hand, aware of the demands, self-critical, working objectively in the undivided interests of the Client, but sensitive to wider issues.

In the quest for explanations of strengths and weaknesses of engineering in Britain, there is much to learn from history, from two episodes in particular. The Greeks developed the first systematic study of science, their Platonic society; the Romans applied this learning but made few attempts to develop it – or even to understand it. The Greeks of the eighteenth century were represented by France, the Romans by Britain. France lacked entrepreneurs; Britain's engineers suffered an excess of competitive enterprise, the high demand compensating, for a while, for the lack of application of a theoretical base of knowledge. Why Britain should have accepted, through this period, the Platonic division between the work of the mind of science and the work of the hand, the successful combination of which is the primary occupation for engineers, is an interesting question. The answer requires an understanding of the collective influence of many interrelated factors of history, geography and philosophy.

In reflecting on the changes observed during a single career in civil engineering, the 'boundary conditions' represent the greatest change. The engineer no longer operates in a self-contained system of providing technical solutions to precisely (or sometimes not too precisely) stated problems. The boundaries are now mobile and porous. The requirement has numerous aspects, the management of risk a feature extending well beyond the engineer's own contribution, solutions complying not only with a Client's requirements but with the effects on others, including

Introduction

posterity. Sustainability is a keyword, a term now never far from a politician's words, if not his thoughts and actions. Politics and engineering have greatly increased their contacts and mutual dependence. Mutual trust lags a long way behind.

It then becomes apparent that the failings of engineering derive sometimes from the engineer not understanding what is politically acceptable, more often from the politician not understanding the limits of practicality, which need to be established by a sound application of science and engineering. Bankers and lawyers, at least sometimes doubtless with the best of intentions, have contributed as much to engineering failures as to engineering success, the partial successes including those achieved despite the penalty of imposed adversarial relationships. It is part of the engineer's function to explain how better to relate the understanding of the engineer to the banker's wish for a positive financial return and the lawyers' positive aspirations, where these exist, for clarity and equity. This is by no means an exhaustive list of the duties of the engineer to external parties. There is increasing contact with agencies for health, safety, working practices and all manner of environmental issues. The function of the engineer therefore needs to be viewed against these many external agencies with whom he needs to communicate clearly, directly or indirectly. Essentially, these represent the contextual features of civil engineering, hence the title.

A book tends to develop its own momentum. It became apparent that statements required support from earlier impressions. There was, therefore, a need for an autobiographical element to provide explanation and to give a degree of justification for views expressed, which derived from experience. Criticism of existing practices is accompanied by recommendations for improvement, sometimes in detail, sometimes by changes in policy that involve others. In Chapter 9, I question many of the political assumptions that condition not only the engineer's task but also the wider effects on the individual. A successful society will increasingly depend on its full use of its engineering capability; I discuss some of the mechanisms whereby this may be achieved.

One particular area merits stress. I express much concern over the extent to which engineers, presented with a technical problem, tend to fly directly to numerical solutions without pausing to understand the physical basis of the problem. Too often, the numerical model does not address the actual conditions or provides output in a form that confuses rather than informs. Almost universally, the apparent precision of a solution bears no relationship to the uncertainty of the data. The ability to make a first estimate on the basis of a sketch and the formulation of equations for simple analysis continues to provide an essential tool to avoid expensive errors and misunderstandings. Education and training of engineers should emphasise the importance of this intermediate stage, exceptions confined only to numerical models which precisely repeat the solution of a familiar conventional problem. Simple analysis indicates the sensitivity of the result to variation of the dominant characteristics. This is not a trivial concern. Recent personal experience has provided strong evidence, if this were needed, of the costly consequences of excessive dependence on a black box, which marks a failing to perform as a professional engineer.

I record with gratitude the unfailing help and advice I have received from the staff of the Library of the Institution of Civil Engineers during the preparation of this book. I particularly appreciate the introductions I have received from the Head Librarian, M. M. Chrimes, to recondite information and to new research which questions received views of earlier authors.

I have no doubt about the problems I recount, nor of the practicality of specific solutions I propose. Where I discuss more general reforms, there are often other options. It is for the reader to consider what these may be. My objective is as much to encourage debate as to proselytise. My belief is that perpetual structural reform of Institutions, excessively competitive (by price) working conditions and project fragmentation, which operate contrary to the interest of the Client, have together combined to discourage the exchange of ideas or reforms of practice that an extended period of change should stimulate. My most general message is that the professional, working in engineering, medicine or education, can help to provide the elements necessary to modify an unworkable free market into an intelligent market. 'Trust' is at the centre of a functioning democracy. Trust is the assurance to be earned by the professional.

If this book stimulates strong criticism, for or against, it will have served the principal part of its purpose. Here my case rests.

1
An historical perspective

1.1 Introduction
So, why should engineers study their own history? Others have, of course, addressed this question and have generally provided three reasons, which, with Morley (1994), we may summarise as: hagiographic; utilitarian; and educational. These are neat enough titles under which to consider the several strands of the arguments, which also need further development against a synthesis of these features.

1.1.1 Hagiographic
The term 'hagiographic' is, of course, overblown for the notion of biographical accounts of the famous engineers, implying an hortatory, uncritical approach. We have literally cast in stone, and set memorials in the most distinguished surroundings (Thomas Telford 1757–1834, George Stephenson 1781–1848, Robert Stephenson 1803–59, Richard Trevithick 1771–1833, Isambard Kingdom Brunel 1806–59, Sir Benjamin Baker 1840–1907, Sir John Wolfe Barry 1836–1918, Sir Charles Parsons 1854–1931, Sir Henry Royce 1863–1933, Lord Kelvin 1824–1907 and Sir William Siemens 1823–83 in Westminster Abbey), to commemorate a number of the best known engineers of the past. While each appears so remarkable individually, we should nevertheless recognise that each responded, perceptively, to the particular opportunities of the time.

Inscribed around the interior of the rotunda of the Institution of Civil Engineers (ICE) and elsewhere in the building appear the same names (with the exception of Robert Stephenson and Sir Henry Royce), together with Smeaton 1724–92, Tredgold 1788–1829, Davy 1778–1829, Maudslay 1771–1831, Dundonald 1775–1860, Hartley

Civil engineering in context

1780–1860, Hodgkinson 1789–1861, Barlow 1776–1862, Faraday 1791–1867, Rankine 1820–72, F. P. Smith 1808–74, Fairbairn 1789–1874, Wheatstone 1802–75, Napier 1791–1876, Penn 1805–78, Maxwell 1831–79, Froude 1810–79, Cooke 1806–79, Whitworth 1803–87, Percy 1817–89, Joule 1818–89, Nasmyth 1808–90, Hawksley 1807–93, Bessemer 1813–98, Rawlinson 1810–98, White 1845–1913, Unwin 1838–1933. To these names, in 2000, were added Arup 1895–1988, Pippard 1891–1969, Pugsley 1903–98, Binnie 1908–89, Gibb 1872–1958, Glanville 1900–76, Baker 1901–85, Freeman 1880–1950. Also the name of Bazalgette (1819–91) was added, having been overlooked in his time.

We should learn to understand the nature of such achievers, something of the strengths – and defects – of character, particularly, for the earlier engineers, often against handicaps of birth, privilege and discouragement. At the present day, there is instant acclaim of mediocrity through a measure of popular mass-hysteria influenced by persuasive agencies (and agents). It could be too easy to imagine those of the past to share something of such transient virtues – until their achievements are set out in sufficient detail to reveal their lasting value. There is much discussion about the suggested low esteem for the engineer. If we do not learn to esteem our most distinguished predecessors, why should we expect to command our lesser share of esteem? We need first, however, to understand their individual situations, their responses to opportunity.

1.1.2 Utilitarian

While utilitarianism is strictly, according to Jeremy Bentham (1748–1832), the notion of according the greatest good for the greatest number, in our context we are concerned with the practical achievements of the engineers of the past with the consequences for society, set against the societies and the economic objectives of the time. Artists and architects need their sponsors – and financiers – for survival; their clients' objectives may be of a personal nature, as were those of the *dilettantes* – the artistic impresarios of their time – or even for personal aggrandisement. Neither attribute should be undervalued in view of the contributions of each to providing many of the greatest treasures of the past. Engineers, on the other hand, need to impress their financiers, as individuals, bankers or shareholders, that there is a perceived need for their artefacts and a prospect of making an investment that will be justified against criteria appropriate for private or public works. Before the days of formal cost/benefit study, a viable return was yet expected on investment.

Other utilitarian aspects of history concern the benefits, in guiding present action, that may be derived from learning about the persistence of our predecessors and from the manner in which they set about solving technical problems in the face of uncertainty.

All prominent engineers will have made mistakes. These errors and their subsequent solutions may well be the most instructive aspects of the utilitarian approach. Generally, we find that each was a pioneer in an unmapped territory; occasional errors of judgement were inevitable while the criteria for success were under constant development and revision. Civil engineers may find that their problems remain remarkably similar to those of the past, the scale and range of options, the severity and magnitude of natural forces the main variables. How did engineers of past ages approach their ambitions? How were resources combined and assembled towards their fulfilment? One feature immediately noticeable is that our predecessors were generally credited at the time with pursuit of the improvement of the conditions of living. They were not confronted by a 'blame' culture with its expensive consequences of litigation. Another feature is that their objectives could be expressed in relatively simple terms, without the complexity of simultaneously attempting to satisfy economic, environmental and social criteria.

1.1.3 Educational

History, well taught, shares with engineering the characteristic of synthesis. In the same way as engineering assembles a concept or project by drawing on applications of science, technology and practical experience, so does history concern the many threads of political, social, religious, scientific and economic thought and development set against a geographical background. The history of engineering is a vital part of this synthesis, too often ignored, as if developments flowed naturally from the current state of society. For example, the colonial expansion of Britain in the nineteenth century was both a cause and a consequence of British engineering capability.

The engineer may set out from an interest in particular technological developments. This may be characterised as the history of capability. But why did the demand occur when and where it did? To what extent were the developments of several technologies mutually interdependent? Such questions can only be answered by some knowledge of the social history of the times.

Civil engineering in context

Dr Robert Legget (Legget 1982) provides a salutary account of the ingratitude to the pioneering engineer as a result of the inconstancy of politicians in response to a shifting historical scene. John By was responsible for the design and construction of the Rideau Canal to allow navigation between the Ottawa River and the Great Lakes, as an alternative to the St Lawrence. This urgent project had been prompted by the 1812–14 skirmish ('Mr Madison's war') between USA and Britain, which revealed the potential vulnerability of Canadian shipping in the St Lawrence to US troops on the right bank. Lieutenant Colonel John By of the Corps of Royal Engineers located the site for the entrance of the Rideau Canal from the Ottawa River in September 1826 and sailed through the completed project in May 1832. The construction had given rise to a settlement lightly named Bytown, now Ottawa. The project, mostly a navigation (i.e. canalised rivers), of total length of around 200 km with 29 km of new canal, had 47 locks and 52 impounding dams. This was a remarkable and unprecedented achievement, delivered 'with despatch', according to instructions. By the time of completion, however, the military objectives were forgotten in Whitehall. Instead of honour, contumely followed in that he was blamed for minor excess of the provision for the work made by H M Treasury, the money having been provided through the Colonial Office, without reference to Parliament, to preserve secrecy. A vindictive Treasury Minute was the immediate cause of his decline. He returned to Britain and died in 1836. There must be few comparable instances of a remarkable engineer being sacrificed by myopic bureaucrats and politicians.

One of the most revealing aspects of the study of engineering history is that of the relationship between theory and practice. To what extent did practical engineering precede a theoretical understanding of the factors involved? How has progress depended upon evolution of theory? The most pragmatic engineers, guided by rules-of-thumb, would gain greater respect from the theorists by demonstrating understanding of the foundations, and the limitations, of the rules-of-thumb that gained acceptance. The academic teacher may, in complementary fashion, acquire a helpful humility in discovering how and where theory followed practice, explaining success and failure retrospectively.

1.2 Uses of history

Our profession, at all levels, needs to cultivate a better understanding of history. At a fundamental level, this is for several practical purposes: to avoid repeating mistakes in our work; to provide guidance in the shaping of our professional institutions; to understand the place for the state in directing policies for engineering; and to foresee how changes in society may influence future acceptance of our projects. An historical context is particularly timely as we move from one millennium to another. The 'new millenniarists' tend to look straight ahead to an unknown destination without apparent awareness that the route towards the destination must depend on the present point of departure and on the contributions of their forebears to attain this point. The Annual Review of the Institution of Civil Engineers of 2000 (ICE 2000a), for example, contains the following remarkable statement: 'We are rapidly changing from dwelling on our historical background to preparing for the future in which we will be working', which contains in a few words one fallacy and one mark of contempt. The fallacy is that the future is independent of the past; the mark of contempt is for the predecessors of the administration of 2000 in supposing that their introspections about the past took precedence over their concerns for the future. Neither of these aspects of the statement provides a satisfactory basis for a healthy perspective. The latter, moreover, is contradicted by much readily available evidence from those who have contributed through the years to the development of our profession of civil engineering and its Institution. St Augustine reminds us that the present slips away imperceptibly such that the future is constantly regenerated as the past. Engineering achievements of the past, whether in works, in understanding or in policies for the profession, were undertaken by those with a perceptive eye for the future. If we choose to ignore the past we unnecessarily handicap the prospects for success in the future. Ignorance of history is a formula for revolution, with all its historical lessons, not for evolution.

One of the objectives of this chapter is to help to correct this misconception and to set planning for the future into a firmer mould. The author would, in political terms, be classed as a radical, having taken part in several of the recent features of reform, some successful, some yet to be fulfilled. History teaches us to find the continuous threads running through our technologies and their applications, our industry and our profession.

A second text for this chapter also derives from the Annual Review for 2000 (ICE 2000a):

> *In the United Kingdom, the nature of our profession has also changed – fewer of the great Smeatons, Brunels and Telfords, but more of the*

5

committed, dedicated civil engineers sustaining the infrastructure of every-day life.

Well, yes, there is more, and more complex, infrastructure to be sustained, i.e. maintained and operated, so there are likely to be more engineers required to be engaged in such activities. Why, though, there should be fewer counterparts of Smeaton *et al.* is less obvious. The great engineers of the distant past, remembered by most at the present day for their heroic contributions, could be counted on the fingers of two hands. We are presumably recalling them for their innovative contributions to civil engineering work and thought. Our increasingly complex multi-disciplinary achievements for the future are going to demand from our leading engineers every bit of resource, imagination and creativity as were displayed by our past role models. Maybe this rather odd notion of compensating for egregious talent by a proliferation of dedicated journeymen stems from lack of understanding of the relationship between inspiration and perspiration manifested by the engineers of the past and, perhaps, from failure to appreciate the different but comparable challenges of the future. Maybe, to be a little more charitable, the statement was poorly expressed, intending to imply that the prowess of engineering teams increasingly takes the place of pioneering individuals. It is dangerous, however, in implying that we no longer need people of vision at a time when this virtue is the commodity in greatest demand. The key capability of the great pioneering engineers of the past was that of judgement, dependent on their personal experience. As technologies develop and proliferate, so, in consequence, does the scope for skilled judgement and leadership increase. Always provided, that is, we develop the skills and do not allow the opportunities to be throttled in bureaucracy and regulation, with its constant resetting of the mousetrap of threatened litigation, which would lead to replacing skilled engineer-managers by administrators. This is the real threat for the future, not the result of excessive backward glances to our historical past. Any statement that implies that we have less need for talent in the future does a disservice to our profession of civil engineering and to engineering in general.

1.3 The Greeks and the Romans

The history of engineering precedes by many centuries the ages of Greek (from say the eighth century BC) and Roman (from say the fifth century BC) civilisations. The construction of many of the

An historical perspective

ancient monuments, including the pyramids of Egypt (say 2700 BC to 1700 BC) and the cities of the Assyrians, dating from around 1500 BC and much wantonly destroyed by Alexander the Great, required the solution to many technical problems and considerable complexity in the organisation of construction. The earliest identifiable engineers applied their skills on the basis of empirical knowledge that they had acquired, often passed on from one to another over a long period. Thus, the Egyptian master builder of 5000 years ago (Viggiani 2001) could apply knowledge specific to particular contexts, such as the flooding of the Nile; he was accorded the reverence and respect due to such a capability. Technology advanced largely by heuristic processes of trial-and-error. Considerably earlier, the working of flint tools, requiring great skill passed down from 'master' to 'apprentice', was, we now know, an achievement of hominids preceding *homo sapiens*. For example, one of our finest sites of early flint tool making, at Boxgrove in West Sussex, is around half a million years old. The discovery, mining, smelting and working of metals developed in comparable manner with the esoteric skills passed down through apprenticeships. Tubal Cain (Genesis, 4:22) is described as 'an instructor in every artifice in brass or iron'.

The use of bitumen in many aspects of building dated from the third millennium BC in Mesopotamian civilisations, also for lining drains, and as an early form of reinforced earth, sandwiched with layers of palm leaves for the ziggurats. Bitumen was obtained from surface suppurations. It is of some interest here to recall that petroleum spirit, centuries later, was first seen as an inconvenient waste product to the manufacture of paraffin wax.

The history of engineering comprises developments in practice and in the underlying theories. The over-simplifying popular view is that the Greeks concentrated on theory, the Romans on practice. Of course, as Humphrey *et al.* (1998) put it: 'the Greeks created and put into use many technological innovations and the Romans spent time on theoretical observations' (p. xvii). However, it is in the relative strengths of the two civilisations that real differences are found, with the Greeks' main fascination with theory, science and philosophy, the Romans in developing the engineering and management skills into artefacts of practical utility.

As I mentioned in my Presidential Address to the ICE in 1977 (Muir Wood 1978), the Greeks made a distinction between work of the mind ($\varepsilon\rho\gamma o\nu$) – ergon – and physical work ($\beta\alpha\nu\underline{\alpha}\theta\upsilon\sigma o\sigma$) – ban$\underline{a}$θusos – the former undertaken by citizens, the latter by menials. This would

7

immediately create a barrier to the development of technology or to the notion of engineering as a worthy pursuit, having regard to the derivation of banaθusos (the root of our pejorative term 'banausic') from the Greek for blacksmith, one who practises manual skills. Finch (1951) reminds us that, at the age of Pericles (c. 490–429 BC) in Athens there were only about 43 000 citizens, against 315 000 others, comprising 115 000 slaves, 28 500 resident aliens and, of course, women. Landels (1978) comments that the Greeks' liking for stability, rest and permanence led to a dislike for change and movement. In consequence, their advances in understanding in matters of statics were perceptive but very vague in relation to dynamics. As a consequence, Plato (c. 428–348 BC) stated that physical objects could not be 'known'. From such attitudes developed the uncritical elevation of the pure sciences above their applications, an attitude which continues to dog a classics-derived education system. While Archimedes (c. 287–212 BC) is of course famous for his contributions to hydrostatics and hydrostatic devices, his ingenuity in wartime machines and stratagems in the defence of the city of Syracuse in Sicily under siege from the Romans in 212 BC is less popularly known. Another name known to us is that of Ctesibus who contributed to the fundamental understanding of hydrostatics in the middle of the third century BC.

Later, an inventive Greek engineer and mathematician, records of whom have survived, was Hero (or Heron), living in Alexandria around AD 65. He was a designer of many devices, more it seems for entertainment than utility. It is suggested (Landels 1978) that he came near to designing a steam engine, further experiments possibly only discouraged by an explosion.

From the viewpoint of ethics and esteem for the professional, Plato has much to offer the engineer. The division between work of the mind and the hand is so reprehensible to the engineer, however, that he tends to ignore and undervalue these positive features. However, the Greeks' attitude to practical skills remains enigmatic. Florman (1976) provides many quotations from Homeric legend of remarkably vivid descriptions of artefacts in metal and many other materials, emphasising not only their contributions to the effective pursuit of love and war but also in the descriptive language which provides admiring, accurate, details of their working and manufacture. The evident ability of the Greeks to pipe water under pressure remains something of a mystery. For example, the water supply to Pergamon crossed two valleys, creating at the lowest point a pressure head of more than 150 m (1.5 MPa). Were the pipes cast in bronze? If so their high intrinsic value could explain why no trace of them survives. The concept of tensile strength of the pipe and of bursting pressure would need some degree of understanding, coupled with pressure testing which would present considerable practical problems prior to commitment to the scheme.

Fundamentally, their engineering achievements were based on applications of their advances in mathematics and in many sciences. Here again, we can only conjecture that, once the degree of development was deemed adequate for purpose, the need for further innovation lost support and the capabilities were passed down through artificers despised by the later philosophers of the Greek civilisation. This disdain has continued strongly to permeate our education system through its excessive dependence on the classics.

The achievements of the Romans, largely for public works, utilities and military purposes, remain widely accessible for inspection across Europe (other than the extreme North) and Asia Minor. The direction of the works was largely in the hands of political leaders, the details left to technicians. As Viggiani (2001) states, 'Romans were not interested in science and technology, but in political and military organisation. They assumed the advanced Greek technology, but in a pre-scientific way. In a few centuries, the understanding of science disappears' (p. 29). This view is echoed by an archaeologist, Richard Stein, writing in *Current Archaeology* (no. 180 p. 512)

> *Roman engineering was mostly a matter of inherited experience, modified over time by trial and error. There is evidence that they heavily over-engineered structures, presumably because they had almost no understanding of the principles of physics involved, and anyway did not have the mathematics to work out the consequences of those principles. Their engineering ability is all the more astonishing for that. But sometimes things went wrong.*

There is much evidence of the Romans' lack of interest in the sciences that underpin engineering, from a reading of Vitruvius, where numeracy is largely confined to dimensions and cost, and including accounts of works that failed to achieve their objectives on account of the neglect of the underpinning principles. Viggiani (2001) describes the Romans' futile attempt to replace the Greeks' piped water supply to Pergamon by a canal. The first attempt in AD 41–51 to drain Lake Fucinus near Rome by a tunnel 5500 m long was unsuccessful through lack of attention to gradient. At the second opening ceremony, once again the hydraulics were misjudged and the celebratory banquet was flooded. With this background, the dome of the Pantheon of AD 125–8, diameter 44 m, with attention to principles of low ratio of weight to strength, using lightweight concrete, the intrados coffered between vault and hoop ribs is, all the more remarkable. Generally their structures were over-engineered, in the absence of understanding of the underlying physics and the mathematics needed in application of science. Reliance on pragmatism impeded the scope for innovation; this remains a general lesson from the history of engineering.

The first stone arch bridge is believed to be Pons Aemilus, dating from 142 BC, but this had been preceded by the true arch in buildings as it replaced the 'false arch' (one that is built up by a series of cantilevered slabs) in the third or fourth century BC, including the 25 m span vault at Ctesiphon.

Whereas the Greeks were predominantly a sea power, with land transport deterred by topography apart from, for example, the tramway of stone slabs (Landels 1978) connecting Corinth to the Saronic Gulf, the Romans depended on a network of roads, transit stations and post-houses. Perhaps the most remarkable survivors from Roman engineering are those concerned with water supply, including cisterns, supply conduits and aqueducts. The water supply for Rome was a particularly remarkable scheme, described at length by Frontinius (Humphrey et al. 1998). There is a known exception to the lack of Greek examples, with the tunnel of Eupalinos (Muir Wood 2000) on the island of Samos dating from the sixth century BC.

1.4 The Industrial Revolution

The decline and fall of the Roman Empire left the cultivation and application of technological skills largely in the hands of Eastern (especially, but not solely, Chinese) civilisation and in the Arab world. The West owes a great debt to the Islamic scholars for preserving Greek learning, which would otherwise mostly have perished, but this story is well beyond the scope of my knowledge or my text. China, as reported by Needham (1971), exported by diffusion many aspects of technology to the West, and during this period undertook major works in building roads and river control works. By design, however, the country was cut off from the rest of the world and there was little incentive in an autocratic society to apply discoveries and inventions in a utilitarian manner. There are many reasons why autocratic rule is inimical to innovation for the benefit of society; we continue to see glaring examples of this truth at the present day.

There were, of course, many remarkable achievements of mediaeval Europe. We immediately think of the surviving monuments, predominantly religious Christian buildings in cathedrals, churches and monastic buildings. We can trace much of the developments in structural form (Mainstone 1998), observing how the mastery of practical problems, such as vaulting over wide areas and the control of lateral forces on the arch, provided incentives for new and ever more daring enterprises. Here, once again, we find the lead through heuristic experience as providing the guide, extrapolating incrementally from past success, learning lessons from failure to apply to limits in daring. The objectives were single-minded, not subject to any deterrence by materialistic cost/benefit studies.

The dome of Brunelleschi (1377–1446) at Florence built between 1420 and 1461 must represent one of the most splendid responses to technical challenge. We know that models were made for the purpose, which doubtless inspired the use of 'stone chains', i.e. blocks of stone used as long stretchers so that they provided tensile strength to combat the circumferential bursting stresses experienced towards the base of the dome. These stone chains were supplemented by iron chains. There are much earlier examples of Byzantine domes, of which St Sophia in Istanbul dating from AD 537 is particularly remarkable, slightly elliptical with a mean diameter of about 31 m and so dimensioned as to avoid dependence on tensile strength of the masonry. The survival of many earthquakes bears testimony to the skill of its Byzantine engineers.

While these few examples demonstrate the natural creativity of the outstanding performers, call them architects or engineers as we will, the limitations are also apparent of progress over many years, dependent on craft skills handed down through the generations. One of the skilled artificers, the millwright, demands particular notice. Monumental architecture needed to pay attention to local circumstances, the terrain, the topography, climate, seismicity and local availability of materials for construction. The millwright had yet more subtly to adapt his skills to the specific characteristics of the site, which itself needed to be selected with great care. In such respects, the millwright, in 'directing the great sources of power in nature' (Royal Charter of Institution of Civil Engineers, 1828), commands special mention as the forerunner of the civil engineer, in many respects, including the essential adaptation of the optimal scheme to the site. It is not surprising, therefore, to find that two of Britain's greatest early engineers James Brindley (1716–72) and John Rennie (1761–1821) gained their first experience working with millwrights.

When we examine more closely the careers of our great engineers we find aspects we may relate to corresponding incidents of the present day. In 1841, for instance (Walker 1841), W. A. Provis described major storm damage to Telford's Menai Straits' road bridge which had caused 175 ft of carriageway to flap in the wind, one third of the suspension rods torn from their bearings and the heads sheared off 3-in diameter bolts as a result of contact between the chains. In the absence of lawyers setting the time-scale, however, traffic was restored within five days.

When elected President in 1845 (after Council indicated a limit to the office of three years (James Walker had served for ten), Sir John Rennie (1794–1874) (Rennie 1846), son of John Rennie (1761–1821), gave a masterly review of the developments in civil engineering since 1724 (the year of Smeaton's birth), his subject loosely defined included all aspects of engineering associated with public works. This was a time for celebrating achievement: 'What triumphant progress have we made? Civil engineer(ing) may be said to embrace everything which can tend to the promotion of the comfort, the happiness and the civilisation of mankind...' (p. 119). This was a remarkable 100-page broad sweep on technological progress. The following year (Rennie 1847) considered prospects for the future. He comments that there was at that time no specific education for engineers, no state support ('Government patronage') but that the engineers of the time had 'attained a degree of skill and eminence in every department... which has been surpassed by no other nation' (p. 27). In these circumstances, it was necessary to be 'careful in the selection of our members' (p. 27), making the services of members indispensable 'by our superior knowledge of the subjects in question, and by the liberality with which we offer them' (p. 29). This period, the middle of the nineteenth century, probably marked the apogee of the technical response of the civil engineering profession in Britain to the industrial revolution.

1.5 The 'Golden Age' and decline

Florman (1976) perceives the years 1850–1950 as the golden age of engineering, but he is seeing the scene from across the Atlantic. In Britain the age was certainly beginning to tarnish by the end of this period and the previous century (1750–1850) may well be seen as more significant in the sheer magnitude and variety of achievement from the starting point of a new profession. An historical perspective can relate such phenomena to wider issues of political and economic

An historical perspective

development, recalling Adam Smith's dampening comment (Buchanan 1989) that 'civilisation, or the improvement of society, has not been the product of human foresight and calculation, but of the natural propensities of economically active man' (p. 2).

Early influences on the different developments of science and engineering in Britain and France may be traced a long way back in history. Roger Bacon (1214?–94), a Franciscan monk of Oxford and Paris, imagined many types of machine for transport by air, land and water and numerous other engineering devices, a long way ahead of the practicability of realisation – and of Leonardo da Vinci (1452–1519). Francis Bacon (1561–1626) recognised the need for scientific theory to be based on experiment, emphasising the particular strength of disproof in advancing knowledge *'major est vis instantiae negativae'* (the strength of negative proof is the greater) (*Novum Organum*, 1620) (Quinn 1980 p. 56) some four hundred years ahead of Karl Popper's comparable theory of scientific advance by falsification. The Royal Society, founded in 1660, had the primary initial purpose of permitting experiments to be performed before the members; its motto *'nullius in verba'* establishes the need for empirical evidence in support of theory.

In France, on the other hand, Cartesian philosophy is based on the belief of Réné Descartes (1596–1656) that the workings of the universe could be fully explained by a unified mathematical system. These different philosophical bases for science at least explain in part the greater British dependence on pragmatism, the greater French dependence on cerebral deduction.

It is instructive to compare early civil engineering in France with that in Britain. In France, major works (prior to the date of the revolution in 1789) were undertaken on behalf of the king under state organisations for the purpose. Work undertaken by Vauban (1633–1707) from 1672 made use of army engineers, the Corps de Génie. The civilian Corps de Ponts et Chaussées was established in 1716 and the Ecole des Ponts et Chaussées (the first civil engineering school) in 1747, probably initially for fairly elementary instruction. To take one example of the approach in France, there was much development in the understanding of hydraulics. Studies in the subject, associated with the Canal du Midi of 1681, and the pumping works at Marly for Versailles, are later exemplified by the 'Architecture Hydraulique' of Belidor in 1738–42, and the theoretical work by Pitot (1695–1771), Bernouilli (1700–82) in Switzerland, and Chézy (1718–98), for example. The state-sponsored canals of the seventeenth century in France, built to a grand scale, did not need to respect commercial profitability, which determined

Civil engineering in context

Britain's utilitarian canals of the following century. Nevertheless, enterprise was stifled by the overbearing state, itself approaching bankruptcy by the time of the Revolution in 1789, and innovation was not a remarkable feature. M. I. Brunel (later Sir Isambard) observed in his youth, at Rouen in 1786, how much more inventive were the imports from Britain. This observation, prompted by his royalist sympathies, caused him to emigrate first to the United States (1793) and later to Britain in 1799. The single notable state-led construction project in Britain in the eighteenth century concerned the Scottish military roads, under the direction of General Wade (1668–1748).

In Britain the situation was totally different from that in France. In France, the Ecole Polytechnique was established in 1794, following the Ecole des Ponts et Chaussées in 1747. The first university schools of engineering in Britain had to wait until 1838. In partial compensation, private enterprise and funding were the keys to activity, spurring the young engineers to work constantly beyond their prior experience, depending largely upon experience gained in success of their own projects and those of others, and on occasional failure. There appeared to be recognition that the engineers were applying their best endeavours so that a momentary setback was not the reflexive subject of litigation. Rennie (1847) states that he 'cannot help speculating on futurity, without previous education or training, without encouragement of... Government patronage...' (p. 27). He continues with warnings that 'we [do not] degenerate into the continental system of centralisation, rendering everything subservient to Government Commissioners' (p. 29). He goes on to emphasise the importance given to the reading of Papers – often extended to a second day – to exchange ideas and to allow others to build upon experience, as a compensation to the absence of organised theoretical knowledge. The engineers of the time appear not anti-academic. They simply did not identify engineering, and here we are concerned with their broad definition of civil engineering, of that time as a theoretical subject suitable for an academic discipline. This was long before the recognition of the benefits to a university from the synthesising influence of a civil engineering department (Legget 1984). Many of the early engineers did, however, recognise the merits of scientific understanding on which practical engineering needed to be founded. W. Cawthorne Unwin in his Presidential Address of 1911 (Unwin 1911) contrasts the lack of technical training, with very little general education, of Brindley, George Stephenson and others with their French contemporaries: Perronet, Gauthery and Navier. In 1866, 93 pupils of Members petitioned

Council of the ICE to consider training of engineers 'so that they might worthily maintain the reputation of their predecessors in face of the strenuous and increasing competition of their rivals on the continent and in America' (Unwin 1911).

In the *Proceedings of the Institution of Civil Engineers* in the 1840s occur frequent references to the dead hand on enterprise perceived as continuing to be cast by the Ponts et Chaussées organisation of France. Robert Stephenson (Stephenson 1856) obviously perceived some of the features of such bureaucracy interfering with the development of the railways. He complains that Railway Law, ostensibly to encourage competition, in fact operated counter-competitively in failing to provide concessions sufficiently attractive to shareholders. He instances a route for a possible railway as a subject for Bills before Parliament in 1845 from 19 competitors. In consequence of such features, he complains that about 25% of the cost of the railway system was disappearing to the law and to conveyancing, with the railways allowed, moreover, to share none of the indirect benefit in increasing the property values of adjacent land-owners. His complaints strike chords with the present day, by such comments as, 'We do not impute to Parliament that it is dishonest – but we impute that it is incompetent' (in relation to the railways) (p. 140) and the statement that 'no railway can be efficiently...conducted without thorough unity among the heads of all the great departments' (p. 144) (Stephenson 1856).

We can regret that the bungled privatisation of the railways in 1993, not only disregarded forward-looking informed advice of the time, but neither did it heed the wise words of experience of Robert Stephenson of 140 years previously.

The dominance of transport in the nineteenth century encouraged a perception of the strong association between the later separated disciplines of railway civil and mechanical engineering. Differentiation was based more on distinction between private manufacturing industry and public transport than between disciplines. In these circumstances, it is not surprising that the Institution of Mechanical Engineers was founded not in London – the essential base at the time for infrastructure and utilities engineering on account of the attention required to Parliamentary Bills – but in Birmingham, nearer the heart of the rapidly developing manufacturing industry, in 1847. For many years after that date, Presidents of the ICE continued to address issues of mechanical engineering associated particularly with transport and infrastructure. This made good sense, helping to avoid, for example, the subsequent insulated departmental bailiwicks of the railways. This later disciplinary division doubtless influenced the fragmented railway structure set up in 1993, obviously so manifest a cause of lack of innovation but this belongs to a subsequent chapter.

While, therefore, we can look back with great admiration at the achievements of the British engineers of the mid-nineteenth century, we can now see that the reliance on learning by experience was, in present jargon, unsustainable. Practice may extrapolate to some degree beyond rational understanding but at this limit the risk increases for experiencing technological freeze or disaster, or for elements of both. Innovation depends on the application of understood principles; pragmatism does not provide these principles. For Britain, therefore, the reason why the golden age of engineering, from 1750, say, was already beginning to tarnish by 1850, may be attributed largely to the result of inadequate understanding of the benefits from engineering education and research.

By the middle of the nineteenth century, already several of the more perceptive were warning of the dangers in the absence of reform. Thus, Lyon Playfair (Playfair 1855) sets out the problem: 'A superabundance of capital may for a time preserve a country from a quick depression, even though it neglects its intellectual training' (p. 49). He then points out that accelerating improvements in transport enabled more rapid spread of competition from overseas into unreformed industries which had previously been protected by inaccessibility. He quotes with approval from Goethe to the effect that nations, as with nature, 'know no pause in ever-increasing movement, development and production – a curse still cleaving to standing still' (p. 52); i.e. innovate or perish. Playfair continues: 'Practice and science must now join together in close alliance, or the former will soon emigrate to other lands' and 'The time is past when practice can go on in the blind and vain confidence of a shallow empiricism, severed from science, like a tree from its roots' (p. 85).

There was thus an evident conflict between the practical engineers, who had made their lasting reputation self-taught in their profession, and the prophets of the time. The prophets not only saw the dangers but also the wider political and economic factors, which tended to obscure the consequences and, thus, reduced any sense of urgency experienced by engineers or statesmen. The attitude to education of the engineer in the nineteenth century in Britain may be represented by a statement of the ICE (Baker 1951) of 1870: 'It is not the custom in England to consider theoretical knowledge as absolutely essential' (p. 262). It is unsurprising, therefore, to find the late dates at which engineering became an academic subject in Britain, i.e. beyond applied mathematics, Newtonian physics or natural philosophy.

It is widely stated that engineering in the eighteenth to mid-nineteenth centuries was not seen as a suitable occupation for the middle classes. Thus, for example, in his *Lives of the Engineers*, Samuel Smiles (Smiles 1874, p. 93) states of Smeaton: 'The educated classes eschewed mechanical callings, which were neither regarded as honourable or remunerative.' Harper (1996) has studied the background of early members of the ICE, based on BDCE (ICE Archives Panel 2000) born between 1770 and 1859 from which he concludes that nearly 80% were from families he classifies as 'upper status' or 'upper or middle class'. Between 1820–30 about 35% of the membership was working as full-time civil engineers (M. M. Chrimes 2004 – and personal communication). The early engineers, such as Brindley and Rennie, learned from millwrights, prior to any distinction between the civil and mechanical disciplines. George Stephenson complained of the elitism of the ICE in 1847, in declining to qualify for membership, when he had already been elected President of the 'highly respectable mechanics' Institute at Birmingham' (to become the Institution of Mechanical Engineers) (Buchanan 1989, p. 79). Such evidence points to civil engineering possibly not sharing the 'banausic' (see Section 1.3) stigma of manufacturing engineering, which was the predominant target of Playfair.

Engineering departments were founded in these British universities and university colleges (later to form universities in their own right) at these dates:

King's College, London, 1838
(Durham, 1838, but failed to survive at this time)
Glasgow, 1840
University College London, 1841 (attempt in 1827–8 failed)
Trinity College, Dublin, 1841
Galway, 1845 (short-lived)
Cork, 1845
Queen's University, Belfast, 1849
Royal School of Mines, London, 1851 (as Government School of Mines)
Owens College Manchester, 1851 (1868 new Chair established)
Edinburgh, 1868 (some teaching previously)
Armstrong College, Newcastle, 1874
Cambridge, 1876 (Robert Willis Jackson taught applied science 1837)
Bristol, 1878

Mason's College, Birmingham, 1881
Dundee, 1882
City and Guilds, London, 1884
Firth College, Sheffield, 1884
Nottingham, 1884
Yorkshire College, Leeds, 1885
Liverpool, 1886
Cardiff, 1890.

By the mid-century, therefore, there were only five engineering departments in the present United Kingdom, of which one was in Ireland. A greater emphasis was given to technical schools and institutes for the artisan. Engineering education was, however, also provided by other establishments:

Royal Military Academy (from eighteenth century)
East India Company College, Addiscombe, 1811
Royal Engineers' Institute Woolwich/Chatham, c. 1820
Putney College of Civil Engineering, 1834 (a private school)
Royal Indian Engineering College, Cooper's Hill, 1871

The pragmatic approach of Britain also extended to the United States, largely as a result of the massive and urgent demand for engineering projects, with little time for subtleties in design or construction. Finch (1951 p. 92) records that the civil engineer's handbook produced by Trautwine in 1872 deliberately excluded any background theory by Rankine, Weisbach and others on the grounds that they represented 'little more than striking instances of how completely the most simple facts may be buried out of sight under heaps of mathematical rubbish'.

During this period, it is instructive to compare the different roots and modes of growth of the different aspects of engineering. If we visualise engineering along a spectrum (Fig. 1.1), we may set civil engineering as the most empirical at one end and electrical (and later electronic) engineering at the other. Electrical engineering developed specifically from science and maintained the conformist mode of advance: research → development → application. For civil engineering, even to the present day, much advance in understanding starts from the observation of a phenomenon in the field leading to investigation and the development of theory which allows extrapolation of the observed phenomenon. Chemical engineering started from a highly empirical base, with the alchemist and the sorcerer dependent for their promises and their displays on closely guarded esoteric formulae confined to

An historical perspective

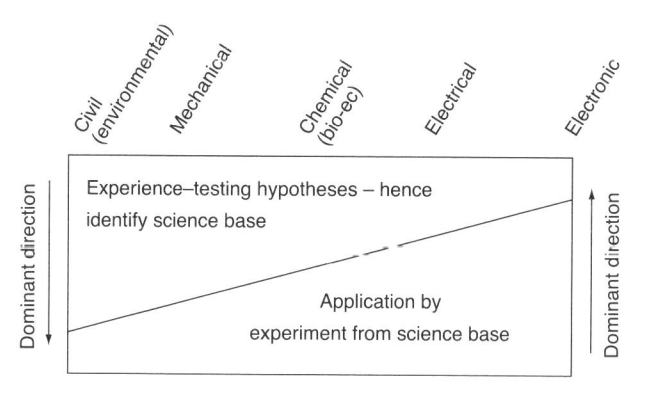

Fig. 1.1 Spectrum of engineering

the knowledge of a few, under conditions of great secrecy. Early chemical-based industries continued to extend comparable practices, with the secrets for success held within craft guilds. Accounts exist of unsuccessful attempts of the Royal Society and of the Royal Society of Arts to break into the scene. From the middle of the nineteenth century, Germany led the way in studying science as the basis for the coal-tar dye industry, leading to the most powerful chemical industry in the world up to the Second World War. In retrospect, therefore, there is much evidence to support the view that, between 1850–70, Britain, having few university-level departments of engineering, and demonstrating contempt for theory of applied science and technology, critically lost the technical lead to Germany.

Mechanical engineering, initially as an indistinguishable part of civil engineering, followed a comparable empirical route. From the first atmospheric engine of Newcomen (1663–1729), powered by the partial vacuum created by cooling low-pressure steam, to James Watt's first rotative engine of 1781, progress was based on a series of pragmatic steps. Watt (1736–1819), an instrument maker, was working in the Department of Natural Philosophy of Glasgow University in 1763 when he appreciated the greater efficiency that might be obtained by separating the condenser from the steam cylinder. This came to him as he was involved in repairing one of Newcomen's engines. By that time, however, he states that his understanding of steam engines was already derived from Desaguliers (1683–1744) and Belidor (1698–c.1761). Watt conducted his own experiments into the relationship between temperature, pressure and unit weight of steam. Later, Watt, now in partnership with Matthew Boulton (1728–1809), obtained a special Act of Parliament to extend the patent under which their engines were made for 25 years from 1775 (Ewing 1926). (Playfair in 1855 observed that a law court 50 years previously had given 'a solemn judgement that Watt had done nothing essential towards' (p. 57) developing the steam engine). Much of the theoretical understanding of steam as a source of power dated from the middle of the nineteenth century with the names of Rankine (1820–72) and William Thomson (1824–1907) – later known as Lord Kelvin – associated with Henry Joule (1818–89) and with Rudolf Clausius (1822–88) in Germany. Here, therefore, the science of thermodynamics served to explain the phenomena of steam engines and their development from about 1840. It is significant that steam engines continued to remain a success story for engineering into the 1890s, particularly around the River Clyde, exploiting the benefit of this marriage between theory and practice.

The origins of electrical engineering, based upon an invisible, intangible, substance, could not have a pragmatic base, beyond elementary experiments with electricity in nature. First it was necessary to discover and measure forces identified by their qualities, to develop basic laws to enable applications to motive power. Electronic engineering and communications were yet more dependent on the 'orthodox' route of scientific curiosity leading to technological applications by way of applied research.

Phenomena of static electricity were observed and described by William Gilbert (1540–1603) in the sixteenth century. He coined the term electron from the Greek ($\eta\lambda\varepsilon\kappa\tau\rho o\nu$) for amber. Benjamin Franklin (1706–90) named the 'opposite' types as positive and negative in 1747. Henry Cavendish (1731–1810) undertook experiments in the 1770s but much of his work remained unpublished for more than a century. Charles Coulomb (1736–1806) studied the effects of charge in the 1780s. The early experiments by Luigi Galvani (1737–98) on contraction of the muscle of a frog's leg (1786) without the use of an electrical generator were explained by Alessandro Volta (1745–1827) as the effect of the potential difference caused by contact between dissimilar metals. The magnetic field associated with the flow of electrical current was studied by Oersted (1777–1851) in 1820 followed by contributions from others, particularly Kelvin (1824–1907). Ohm's law dates from 1827. Michael Faraday (1791–1867) discovered electro-magnetic induction in 1831 and made the first simple electric motor. The practical solution of a direct current generator is attributed to Charles Siemens (1823–83) around 1870. There is a comparable history of electro-chemistry. Once again is illustrated the 'orthodox' route from scientific discovery → experiment and development of theory → practical application.

At the cutting edge of electronic engineering, new devices and applications depend intimately on developments in the underlying science. The technology is science-led to the degree that the distinction here between physics and technology is thoroughly blurred, with patents launched from the laboratory. Here we may contrast civil engineering, whose basic laws are those that have been recognised in nature, if not quantifiable, over the centuries. The protection of a device, of a concrete shell dangling on a pile, for absorbing energy of ships coming alongside, the Baker's Bell, was held by its detractors to amount to attempting to patent the law of gravity.

2

Recent historical perspective

2.1 Dominant features

Chapter 1 discussed some of the principal factors, interpreted in a broad sense, providing the background to the current technical capabilities of the civil engineer. A brief account was also given of the procedures for commissioning engineering services and the means for procuring their achievement. Essentially, by the second half of the nineteenth century, two different basic systems for procurement had developed. In many European countries, work was commissioned by the state, with the lead provided from France, where the Revolution transferred the client from the Crown to national bodies such as the Ponts et Chaussées organisation. In Britain and the US, however, a totally different tradition had developed, with works for transport such as canals, railways and roads, and later water supply, financed by companies supported by private capital. The New River Water Company, bringing water to London, was the first water supply company to be operated as a Trust. Turnpike roads, also several port and navigation authorities, were established as trusts, financed from dues levied on the users of the facilities by the trustees.

These different traditions led to an increasing separation between design (of the permanent work) and construction, the division perpetuated while the latter gradually developed particular skills beyond those of the artisan tradesman. Design was accomplished by groups of professional engineers, employed by the State or, in countries following the example of Britain, by private firms which, in the course of time, assumed the characteristics of consulting engineers, where the Principal as Engineer, or Principals as Partners, undertook the responsibility for the work. Initially such Partnerships often developed around an engineer who had earlier held a responsible position in an organisation which itself had commissioned major engineering works. The Engineer undertook the planning of the work, in conjunction with the Client, commissioned studies considered necessary to underpin the work, designed the work, prepared contract documents, engaged contractors on behalf of the Client and supervised the construction.

The great British engineers of the late eighteenth and early nineteenth centuries had a variety of different approaches to the undertaking of their projects (Barnes 2000). Occasionally this was by what we would nowadays know as 'design and build' where the engineer required only the provision of materials and labour, himself organising the undertaking of the work. One instance is that of James Brindley (1716–72) in the construction of the first Harecastle Tunnel on the Trent and Mersey Canal, another that of M. I. (Sir Isambard) Brunel in the construction of the Thames Tunnel. John Smeaton (1724–92) developed the procedure of the supervision of work undertaken by contractor, appointed on merit, through the Resident Engineer, in certain respects the forerunner of what became known as the 'traditional method' in Britain (see Section 2.2). The most successful projects were based on the selection of contractor on the basis of prior performance, an equitable form of payment and a co-operative form of management, as practised by John Rennie (1761–1821), Thomas Telford (1757–1834) and Joseph Locke (1805–60).

For a direct comparison, Brindley's first Harecastle Tunnel, constructed by direct labour, required 11 years (1766–77) for its construction, delaying the opening of the canal by two years and, later, providing evidence of poor quality of workmanship. Telford's second Harecastle Tunnel, entailing the employment of his chosen contractor, Pritchard and Hoof, was successfully constructed between 1824 and 1827. John Rennie used large firms in serial construction of projects, particularly with the contractors Jolliffe and Banks (Banks the technical director, Jolliffe – an ordained minister of the Church of England – the commercial one).

The organisations contracting to execute the work were developing in a parallel and complementary manner, from providers of men and materials contracting to undertake work in a traditional, labour-intensive manner, to the management of multi-functional organisations, with access to purpose-designed plant, techniques and equipment. The capability also developed to design increasingly complex procedures and temporary works. Thomas Brassey (1805–70) undertook the first BOT (build–operate–transfer) projects paid for by shares in the company formed for the purpose (which would, at the present day, be called a Special Purpose Vehicle).

Civil engineering in context

In the early days of the period under review in this chapter, from say 1900, the undertaking of a civil engineering project could be described as a linear process of Planning → (Project-specific studies + design) → Construction. There were departures from this route by imaginative engineers engaged in particularly unprecedented forms of engineering. Many years earlier, for example, for the Thames Tunnel (1825–41), M. I. Brunel recognised the need to take the principal decisions not only in relation to the permanent work but also to the design and techniques of construction, evolving as the work proceeded.

Sir Joseph Bazalgette (1819–91) may be considered to be the real progenitor of the 'traditional method' of contract management in Britain in recognition of the standard form for London's Metropolitan Board of Works, used among other projects for the construction of London's main drainage in the 1860s, and serving as a principal model for more than 100 years. The First Edition of the *ICE Conditions*, the first universal standard for civil engineering, was only published in 1945.

2.2 The so-called 'traditional method' of project procurement and management

The predominant system in the UK which developed over around 150 years from the middle of the nineteenth century depended upon a professional relationship between the Employer (or Client of the Engineer) and Engineer, and upon a subsequent commercial relationship between the Employer and the Contractor. These were achieved through the means of an Agreement between Employer and Engineer, and a Contract between Employer and Contractor, drawn up by the Engineer, including, from 1945 onwards, usually the ICE Conditions of Contract, which designated the 'Engineer' as effectively the independent controller of the Contract.

In essence, the traditional method designated three functions for the Engineer: the design of the Works; the supervision of construction and the measurement of the works for payment; the service as arbiter in first instance for any matter of interpretation of the Contract. The principal virtue was that of a project controlled by an engineer familiar with the project and with command of the processes of design and construction appropriate for the work, including the selection of a competent Contractor. The Engineer would rapidly recognise a circumstance which was unexpected by his design. In principle, therefore, and often in practice, such an arrangement provided a good chance for an optimal engineering solution for the Employer, including variation for the unexpected, with fair reward for the Contractor. He did not need to allow substantial contingencies for features over which he had no control and no basis for estimation. Essentially, the likelihood of some degree of uncertainty was recognised in natural circumstances, or in quality of work, best to meet the demands of the Engineer's design for the project, undertaken on behalf of his Client.

Contrary to much uninformed opinion, the traditional method worked well when the essential conditions for it to work were honoured:

1. The Engineer is fully conversant with the Employer's objectives and competent to interpret these into an optimal project.
2. The Engineer is appointed on merit for the particular nature of the work.
3. The Engineer participated in the preparation of his own terms of reference with fair terms of reimbursement (primarily, to ensure their appropriateness for the achievement of the Employer's objectives).
4. The Engineer is given overall responsibility for project planning, design and studies appropriate to support the project (e.g. site investigation).
5. Tendering contractors and the appointed Contractor are selected and assessed in relation to their technical competence, means and probity in relation to the nature of the project.
6. The Engineer kept the Employer fully informed on the nature of uncertainties in relation to the project. In return the Employer recognised that perfection in engineering does not exist.

The conditions set out above imply a professional relationship of trust between Employer and Engineer, recognising that for success the Engineer must exercise an objective and independent control of the Contract. The Engineer must also deserve full trust from the Contractor particularly, but not exclusively, in relation to the unexpected. The Engineer is primarily concerned with achieving the Employer's objectives through the optimal management of resources. There are many Employers, Engineers and Contractors who have experience of a high degree of success in relation to Contracts undertaken through the 'traditional method'. Where failures have occurred, one or more of these factors will have intervened:

(a) The Engineer appointed in competition entailing features other than quality. At the extreme, the appointment is made under some such name as 'Design Contractor' which establishes that the project will be managed by an administrator with inadequate understanding of the engineering issues. Apart from the lack of satisfactory means of competitive appointment to undertake tasks incapable of precise definition, the Engineer then has no hand in the preparation of the nature and breadth of his own Terms of Reference, vital if the Employer's objectives are to be achieved optimally. The Engineer is appointed to undertake minimum duties specified by the Employer, not to undertake duties essential to the success of the Project. A way around this problem, if statutory law demands a competitive appointment, may be to extend an initial appointment for preliminary project definition, at a later time when definition of duties is possible.

(b) The placing of a discontinuity, possibly on purpose, between the function of design and that of construction. Project Finance Banks, including The World Bank, go so far as to disqualify the project designer from engaging in supervision of the same project, at one bound impairing significantly the prospects for success.

(c) Encroachment on the powers of the Engineer, perhaps by making his decision subject to the agreement of the Employer, or by removing terms from the Conditions of Contract essential for its fair administration, e.g. those for 'unforeseeable conditions'. In this way, the Engineer's powers to rule on uncertainty are removed, or the ability to contribute, for example by revised design, to overcome an unexpected problem.

(d) Fragmentation of responsibility for other aspects, e.g. ground studies, so that these are no longer related to the specific scheme(s) for design and construction. An essential feature of a successful project must be that of purposeful engineering continuity providing synthesis across the contributory elements.

(e) Either Party to the Contract has a conditioned reflex to reach for a lawyer as soon as a problem is encountered, an attitude fostered by an initial display of risk aversion by the Employer.

Here it is relevant to inject a note on the concept of the Intelligent Market, a concept discussed by Muir Wood and Duffy (1991). The Intelligent Market recognises the existence of values in market transactions that transcend costs based on price. The concept is further discussed in Section 5.1. The mark of 'intelligence' in this respect is the ability to perceive the optimal procurement strategy for a construction project overall, not the expectation of least cost by means of engaging the individual parts separately and without consideration for adequate synergy. Essentially, the quest throughout the process of optimal project procurement for each decision is for value, not cost.

Where professional relationships are concerned, there needs to be a recognition that unbridled competition offers a direct threat to achieving value for money. The essential place of professionalism is argued in Chapter 3. Here it is only necessary to observe that it lies at the heart of the 'traditional method'. Criticism or meddling with the traditional method without regard for this fact is unhelpful and has had a damaging retrograde effect on construction in Britain and elsewhere, by preventing integration and the long-term investment essential for the innovative outlook of successful engineering.

In summary, the 'traditional method' has become distorted as a result of a misguided attempt to apply to it over-simple rules of the market place, suitable for simple transactions but which are not relevant, and are, in fact, destructive, to professional relationships. These distortions have caused the traditional method to become dysfunctional, or at the best to function sub-optimally. In consequence, the cry from the uninformed is to discard the whole, the baby with the bath-water. The proper remedy is a change, or simply to restore the cleansing properties, of the bath-water.

2.3 What went wrong with the ICE Conditions?

We need at the outset to differentiate between popular perception and fact as to the reasons for the ICE Conditions (and the similarly structured Fédération Internationale des Ingénieurs Conseils (FIDIC) Conditions) receiving a recent adverse press. As one who has used such Conditions to achieve success for all Parties to a Project, some of which are summarised in later chapters, I feel qualified to present the case for the defence. The criteria for success should comprise: a Client whose requirements have been understood and fulfilled; a Contractor who has been adequately reimbursed for a job well done; an Engineer who has fully understood the Client's needs – and has probably contributed to the Client's perceptions of what these should be – has applied competence and creativity to a well-engineered project; the rarity of unresolved dispute and litigation.

The ICE Conditions assume that the Engineer has a thorough grasp of the requirements by the Client for the project, including conditions for its operation and management, recognising that the costs incurred during the lifetime of the project may well exceed the costs of construction. For a project of a nature unfamiliar to the Client, the Engineer may well contribute directly or indirectly to this knowhow. The next essential is that the Engineer possesses an adequate degree of competence throughout the range of engineering appropriate to the project – or knows where to obtain suitable reinforcement from others to compensate deficiency – to perceive the optimal overall solution, which may well entail innovation and the adoption of specific means for reducing uncertainty which might develop to risk.

Traditionally, the Engineer needed to understand and respect his several different but interrelated roles: designer of the Works as agent to the Employer; as supervisor of the Works to ensure that specified requirements are achieved; as interpreter of the Contract in a capacity independent of the Employer. The essential environment for undertaking these otherwise potentially conflicting duties is that of mutual trust with Employer and Contractor. The Engineer, in interpreting the Contract, will be guided by the reasonable expectation of the Contractor at the time of Tender. If any Party introduces lawyers to the discussions, in which the Engineer will be engaged prior to making decisions, recondite matters, far from those considered by the players at the relevant time, will make the Engineer's functions unreasonably difficult. To maintain trust he must, nevertheless, resist departure from the maintenance of reasonableness. The performance of the function of adjudicator becomes simpler and more equitable where the Engineer has foreseen risks, that may develop from project uncertainty whose extent and cost are unforeseeable, and which should be excluded from the Contractor's responsibilities. In later years, say from the late 1970s, restrictions on the independence of the Engineer threatened the equilibrium of the ICE Conditions. It deserves note that a form of contract favoured by Government Departments at this time, GC/Works 1, curtailed the powers of the Engineer to give instructions to the Contractor or to vary the design of the Works, thus severing the link between design of the works and design of construction.

Another element of criticism of the ICE Conditions arose from engineers of inadequate competence being appointed, at the time of major expansion of construction in the 1970s, as Engineer. It was too easy, in the absence of an over-riding professional motivation, to pass on responsibilities, properly the Engineer's, to the Contractor, and to allocate risk improperly and without adequate thought. All such factors contributed to expensive and unsatisfactory projects, with increasing numbers of lawyers becoming involved in the one remaining lucrative aspect of construction.

The ICE Conditions represent no universal panacea to the greater and increasing complexities of the scene of construction. They continue to provide a good model for appropriate circumstances, particularly where the dominant innovation stems from the design of the project or of significant elements, coupled with techniques of construction adapted to the innovation. A competent engineer should be capable of recognising where these conditions exist. The principles of professional motivation, of mutual co-operation and competence, of indivisibility between engineering and management remain paramount and need to be demonstrated by any alternative model.

2.4 Recent reports affecting construction

There have been a number of reports in recent years with important but not wholly beneficial consequences on construction. In the following paragraphs, only those of particular significance have been selected for comment. Numerous reports unmentioned have been published as guidance on the application of major reports or in extension of their scope.

2.4.1 Banwell Report

The report of the Committee chaired by Sir Harold Banwell (Banwell 1964) was primarily concerned with improvement of the general standard of contract practices, using models from what were considered the best practices of the time. Recommendations were made for a limited number of tenderers selected through a one- or two-stage process. There was support for serial contracting, i.e. the appointment for one contract leading on to another, an early form of partnering. No action followed from the recommendation of a single standard form of contract for all civil engineering and building.

2.4.2 Latham Report

The Latham Report (Latham 1994) is a mixture of several sound precepts, solid particles in a medium of diluted analysis and not entirely coherent reasoning. The writers demonstrate more familiarity with small-scale building and its problems than with the rather different requirements, including examples of the more notable successes of construction or civil engineering. Since this has been a seminal report from which much else has derived in succeeding years, a little time is

Civil engineering in context

justified in a critical assessment. Several statements upon which the report depends require much qualification in relation to civil engineering.

The principal terms of reference were 'to consider current procurement and contractual arrangements and current roles, responsibilities and performances of the participants, including the client' (App. 1, p. 113). By implication, this is conceived as an administrative task, without regard for the understanding of the professional, or even the technical elements, vital to project success, upon which an appropriate contractual environment should be constructed.

Latham rightly identifies (Cl. 1.11) much of the problem of project procurement of the time as having derived from the dismantling of technical competence of the public service as a result of government policy in Britain (mostly under the Thatcher Government 1979–90). The attack at this period on entrenched bureaucracy, initially justified but subsequently pursued excessively in a doctrinal manner, had resulted in the Public Client finding itself with inadequate technical resources to safeguard its proper interests, an essential element of the Intelligent Market mentioned in Section 2.2 above. Not only does Latham not attempt any more subtle distinctions between the particular features of different classes of work, but nor is there differentiation between house building and construction. To illustrate such an important feature for construction, it is only necessary to consider the extent to which the design may be intimately related to particular aspects of the process of construction; another example will concern work with a distinctive element of innovation. The level of research is described as inadequate (Cl. 7.31) but not why nor what one might do with it. Nor, for example, is there any suggestion as to how the planning and recompense of construction activity might encourage work on development, particularly for such work to be undertaken during slack periods (Cl. 2.5). Without such a policy, the demonstrable weaknesses in technical development of British construction industry will persist, without funding from accumulated profits when demand is slack, without surplus resources when demand recovers. Demonstrably, this is not a technical report, so the significance of innovation receives no mention. Yet, this has been at the centre of the problem for decades: the industry engaged at least cost, working to narrow margins, with little investment in consequence in innovation and training.

The report addresses projects which are based on commercial rather than professional relationships. A commercial transaction depends on the ability to define, within adequate precision, what is to occur and the circumstances of the occurrence. As one example of many of this prejudice, notwithstanding the available overwhelming evidence of the successful operation of the 'traditional method' when properly operated, this statement is made (Cl. 3.6 and again in Cl. 3.15 as if repetition made it the more credible): 'Any client who wants external advice over project strategy and need definition should only retain an adviser on the express understanding that the role will terminate once the decision has been formulated on whether or not to proceed.' There have been several proposals for providing a prospective Client with adequate competence (in commissioning a project and in its operation) to proceed with a project, e.g. an initiative by consulting engineers in Scotland and the notion of the Surrogate Operator (Muir Wood 1995), but deliberate disruption of continuity between project strategy and its execution is unhelpful. This is precisely the point at which the inexperienced Client is liable to stumble. The implication of Cl. 3.7 is that the Client should assess risk internally to devise a contract strategy of how much risk to accept, without a knowledgeable adviser to establish a sound basis for risk-sharing. In the absence of advice which is related to the nature of the project and the appropriate 'ownership' of risk, a policy of deflecting risk onto others is likely to have the effect of increasing cost, increasing risk overall and impeding risk control. Without a wise evaluation of the circumstances, an apparent low-risk strategy is likely to rebound, enhance risk and ultimately increase the Client's exposure. At a stroke, without consideration of the long tradition of the ability of competent consultancy to identify with the Client's interest, an effective means of assuring continuity in working in construction is jettisoned without attempt at replacement of the positive – and frequently vital – features of continuity. Purposeful continuity is usually at the heart of a major successful project.

Latham recognises the vital part to be played by the public client in promoting good design, without showing a clear appreciation of the breadth of scope of the design process. The most vital decisions concern the procurement route. Cl. 3.7 states that this decision should precede the preparation of an outline project brief. While admitting that inexperienced Clients will require advice at this point, the advice is given that 'there are a number of publications which can assist'. Each such statement, unqualified as to the nature of the project, exposes a lack of understanding of the essential integration for success between administration and the engineering, a failure to understand the nature of major construction projects and an apparent instinctive suspicion of those professionals capable of providing continuity of guidance through the vital passages of a developing project.

The bogy is the portrayal of 'traditional construction' as inevitably entailing lack of co-ordination between design (of the project) and construction, without qualification as to veracity or to search for explanations, as suggested in Section 2.2 of this chapter, where this has occurred. (Contrary examples from my direct personal experience are also provided in Chapter 6.) Latham provides advice on discharging the designer after certification that the design is complete. Once the designer is discharged, there is little prospect of varying the design to suit an advantageous form of construction or to advise on how to meet an unforeseen change in circumstances or in the Client's requirements. 'Innovative construction' is offered as an alternative to 'traditional construction' with the prospect of construction management with hands-on involvement. The translation of the Owner's 'business case' into the most appropriate engineering response is possibly the most skilled part of the Engineer's function. If departure from 'traditional construction' is to be advocated, great new exposure of the Client to risk occurs in the absence of a well thought through alternative means of providing continuity.

The document contains several positive advocacies of the practice of 'signing-off' by those who have completed particular tasks, including checking of proposals. This function could also contribute to project discontinuity, by discouraging the revisiting of a particular decision which may be affected as a result of iteration. Too many projects to which engineers are attached in a position of auditor require the 'signing-off' of individual elements which, to make sense, should be viewed integrally. For example, design of elements are submitted for checking and 'signing-off' without evidence of the particular requirements from construction techniques needed to ensure that any special demand of the design is honoured – or even have a chance to be honoured if given special attention. The auditor may not even have detailed knowledge that an appropriate relationship between designer and contractor exists for such assurance. Another instance arises where a general issue, such as fire safety, requires 'signing-off' of individual elements separately, whereas the adequacy of all forms of safety can only be assessed with an overall knowledge of all contributory circumstances of design, construction and operation. Of course, there needs to be a record that a task has been undertaken but this should be achieved in such a manner to contribute to the whole, avoiding confusion in responsibility, rather than as an agent to fragmentation. In essence there needs to be a hierarchical design of any vetting process, with the higher level operation ensuring that any change overall is transmitted to the lower level to ensure that prior 'signing off' is not invalidated. The higher level body takes responsibility overall. In essence, the process corresponds to any other system dealing with risk, as described subsequently.

Latham presents (Cl. 5.8) surveys on attitudes to standard forms of contract but it is not always possible to identify the reasons for success and failure which, one might suppose, alone would justify such a survey. Contracts based on separation between design and construction are then condemned as unacceptable (Cl. 5.17), also the combined function of Engineer as designer and adjudicator of the project.

Towards the end of the Latham Report, several recommendations point in the right direction. Thus, for example, there should be a register of consultants (Cl. 6.11) 'to demonstrate some appropriate professional and managerial skills, resources and professional indemnity insurance'. There is acceptance (Cl. 6.14) that the 'lead designer' may act as Project Manager, although this could contradict what has gone before if the Project Manager is to have control of the nature of the project, unless the lead designer appears after the project concept has been determined. Clause 6.36 states that all tenders should be evaluated for quality, cost-in-use (i.e. full life costs), out-turn and past performance of tenderer. Clause 6.47 discerns merit in the notion of partnering by way of competitive contract without guidance as to the circumstances or conditions for securing such an arrangement, particularly in the absence of the designer. The Construction Design Management (CDM) team may be involved in the project in other capacities; particularly where the designer is prevented from contributing through the construction phase, there is a strong risk, however, that CDM undertaken by those lacking design judgement may obstruct desirable innovation.

Clauses 7.31 and 7.42 describe the inadequacies of research and of application of Quality Assurance (QA) through BS5750. Finally, complete confidence is placed in the New Engineering Contract (see Section 2.5 below), proposals for which were still being developed at the time of the Latham Report (1994) and which had yet to be tested in practice.

Latham recommends the replacement of the Engineer by the Adjudicator, with the potential Adjudicator named in the Contract (Cl. 9.7). The relative merits of the several forms of dispute resolution are discussed in Chapter 4 of this book; here it is only pertinent to point to the weakness of the adjudicator in being unable to anticipate

Civil engineering in context

problems and thereby contribute to their solution by possible design changes. The designer, fully acquainted with the unfolding project, can achieve economy by heading off unforeseen circumstances, which would otherwise become the cause of variation and possible dispute.

In Clause 9.1 occurs the observation: 'During the last 50 years much of the US construction environment has been degraded from one of a positive relationship between all members of the project team to a contest consumed in fault finding and defensiveness which results in litigation.' This merits a personal note. In 1977 I was concerned with the preparation of a document on risk-sharing in tunnelling, subsequently published by the Construction Industry Research and Information Association (CIRIA 1978). As a consequence I was approached by a delegation from the US who wished to meet the organisation of British construction lawyers; at that time I knew individual British construction lawyers but they were then marginal to the construction industry and had nothing like the power and influence of their US counterparts. It further merits mention that the greatest single cause of litigation and allied obstruction to innovation in construction in the US has flowed from the discontinuity, a barrier, between design (up to Tender) and construction. The appointed Contractor had to accept full overall responsibility, a practice only now in the course of change.

2.4.3 Egan Report

Although not in strict time sequence, it is well to consider the report *Rethinking Construction* (Egan 1998) immediately after the Latham Report with which it merits comparison. Egan identifies the main weaknesses of the construction industry as low profitability, inadequate expenditure on research and development, inadequate capital and training, coupled with Clients who confuse value with cost. The five major themes for improvement entail:

- committed leadership,
- focus on the interests of the customer (Client),
- integrated processes and teams,
- quality-driven agenda,
- commitment to people.

The most sweeping changes, for example for partnering which dispenses with competitive tendering (on price) and relies instead on performance measurement, have greater relevance for major Clients and major projects. The fundamental messages – of co-operation, not

Recent historical perspective

blame; of continuity through all processes, not fragmentation – have general application. Egan recommends radical change; many of his recommendations have been put to the test and found to deliver savings in cost and time comparable to his targets. Egan also advocates as much as possible of construction being undertaken under factory conditions. Here again, the French bored section of the Channel Tunnel, the deck elements of the Second Severn Crossing, much of Terminal 5 at Heathrow Airport and the tunnel elements for the Øresund Link demonstrate the benefits of such an approach.

Egan emphasises the merits of standardisation, examples from manufacturing industry leading to 'lean thinking', and benchmarking, which are all relevant but bring less benefit to the 'one-off' project.

Perhaps the most striking feature of the Egan Report is the insistence that the project is driven by the needs of the Client as customer. This vital thought has been developed by others, including Blockley and Godfrey (2000) who have taken the title of their book *Doing it Differently* directly from Egan.

2.5 The New Engineering Contract

The New Engineering Contract (NEC)(ICE 1993) was a document to simplify and improve the traditional contractual practices of the UK. In the present context, the NEC is also identified with the subsequent editions and with the several explanatory reports and guidance notes issued at intervals since 1994. Many of the issues identified by Latham (Section 2.4.2) were incorporated into the NEC. A second, fully 'Latham-compatible' edition of the NEC was published in 1995.

The NEC is deliberately written in simple language, an attractive feature to engineers if not to lawyers, and provides for different forms of reimbursement, by lump sum, through target contract or measured work, including also for design-and-build. The potential success of the document is demonstrated by the recent complaint of a lawyer of the lack of opportunity to test the legal construction of the document (Philip Capper, *NCE* 16.1.03) – long may this situation last! The documents of the NEC are constantly urging co-operation and teamwork, but fail to identify the conditions to provide an appropriate environment. It is this issue which mainly concerns the context of this chapter.

The principal defect of the NEC appears to be the replacement of the Engineer, parcelling his duties around the Supervisor, designer and Project Manager, with just a hint that these may be recombined. In simple commercial logic, the duties of the Engineer of the ICE

Civil engineering in context

Conditions are potentially conflicting, in acting as Agent to the Client, as designer of the works and as regulator of the Contract between Client and Contractor. If the practice is reassessed in a professional environment, this is not irreconcilable. It is the sort of price – and once again I emphasise that there are other ways of achieving the same objectives – to pay for continuity and informed understanding of the project features which will help to resolve the most significant uncertainties. The essence lies in the notion of purposeful continuity throughout the project. The NEC acknowledges that unforeseen events will happen and has early warning procedures. It then treats these as 'compensation events' (a 'zero-sum + costs' occasion). As a professional engineer, my prime objective is to foresee and prevent occurrence in the prime place. The designer as controller is able to foresee potential departures from expectation and head off the consequences before they occur, thereby achieving a positive gain for all concerned. There is also a wide, if immeasurable, benefit from the co-operative attitudes that such anticipation engenders.

Teamwork depends principally upon a perception of shared self-interest in a particular course of action. For engineers, a powerful influence towards teamwork lies in the perception of its contribution to a well-engineered project. This attitude may be expressed as an aspect of a professional outlook, widely emphasised elsewhere in this book.

It is in the manner of overcoming the unforeseen, but usually foreseeable, problems that the main defect is seen to arise from departure from the 'traditional' method (e.g. a contract based on the ICE Conditions) (see Section 2.3 above). Foresight in such respects and anticipation (i.e. taking action in advance) of the need for change before the problem occurs is likely to be achieved by an Engineer familiar with the basis of the design of the project, in the systems and means adopted for its undertaking. Such a capability will be possessed by an Engineer who:

- has undertaken the design and planning,
- is fully familiar with the circumstances of the project, including most, particularly the Client's interests,
- has the authority to require variations in the manner in which the work is tackled.

These are attributes of the Engineer of tradition; the NEC, as described above, subdivides the functions into three parts, namely: Project Manager; Supervisor and designer. The functions of the designer are misrepresented as confined to the preparation of a 'design', considered as a single product rather than part of a process. The role of the designer is thus confined to developing the project design to meet the Employer's objectives to the point where tenders for construction are to be invited (or the preparation of a performance specification for design-and-build). Chapter 4 of this book addresses this problem and searches for solutions, while Chapter 5 establishes the substantial merits of 'purposeful continuity' in the handling of a project.

In summary, the NEC is a suitable vehicle for the application of good practice, but it needs to be used by those with enlightenment if the potential is to be realised. The intentions stated throughout the documents are unexceptionable. One essential feature is an understanding, shared between the parties to the contract, of the new distribution of duties. If not, the requisite co-operative attitudes will not occur. An observation made by Sir Charles Inglis in 1941 (Inglis 1941) continues to remain apposite: 'In many organisations there exists at the top an upper stratosphere of conspicuous ability and mature judgement and at the bottom there is found a layer of youthful enthusiasm and fresh ideas,..., in between there is apt to be a stratum of high resistance and poor permeability...of those perpetuating in a painstaking manner the mistakes of their predecessors'. At times of change, particularly, this warning should be heeded. The intentions of the NEC may be, and are, thwarted by middle management continuing to work in a 'traditional' manner, oblivious of new responsibilities and attitudes.

2.6 The FIDIC Suite of *Contract Documents* (1999)

Does the 1999 suite of FIDIC *Contract Documents* (FIDIC 1999) help or hinder prospects for successful projects? The Silver document for EPC/Turnkey Projects has limited provision for interplay between the Parties of Employer and Contractor, through instructions which may, or may not (Silver FIDIC Cl. 3.4), give rise to variation. For such Contracts, by far the greater need for interactions occur within the Contractor's organisation; EPC (engineering, procurement and construction) type projects continue to give rise to problems on this account. My present concern lies with the red (construction) and yellow (plant and design-build) but principally with the former, which I refer to as 'red FIDIC', and to which the references relate. I concentrate on:

(a) provisions for continuity in planning the overall development of the project,
(b) encouragement for constructive relationships between the Parties,
(c) provisions for dealing with uncertainty.

Civil engineering in context

2.6.1 Provisions for continuity
Issues of continuity generally lie beyond the scope of red FIDIC. Nevertheless, two points should be stressed, for a Code of Good Practice that might be expected to accompany such a document:

(a) The Engineer, if he also designed the project, will be in a position to understand the essential features of the Contract and to perceive, to the benefit of the Employer, where conditions depart from expectations of the design, and thus avoid consequential complications.
(b) Fragmentation of the coupled features of planning, studies and design (Fig. 2.1) should be avoided by early appointment of the Engineer as designer.

2.6.2 Relationships between the Parties
FIDIC (1987), the latest edition of the precedent document, represents the Engineer as an independent agent in relation to the Contract (Cl. 2.1) except in so far as he might need prior approval of the Employer, as prescribed. By red FIDIC, Cl. 3.5, the Engineer is required to 'make a fair determination in accordance with the Contract, taking due regard of all relevant circumstances'. This wording appears to limit the Engineer's reliance on professional judgement. Traditionally, the

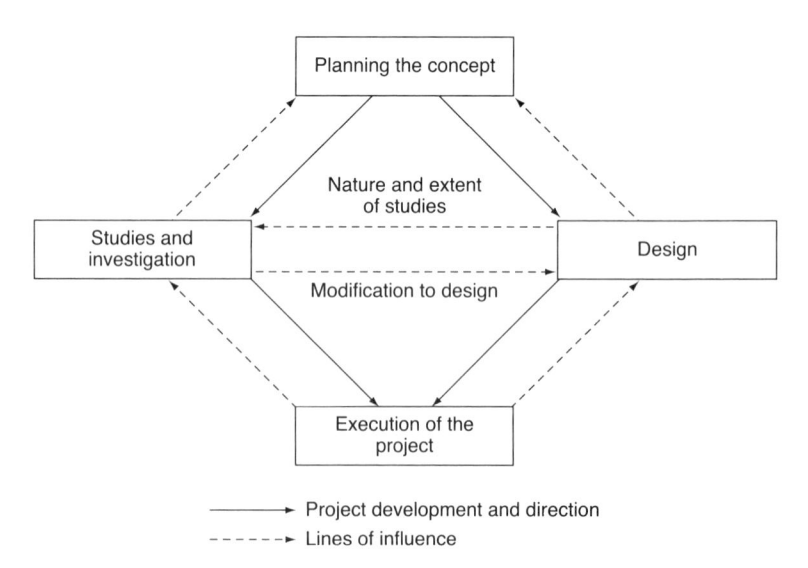

Fig. 2.1 Coupled features of planning, studies, design and execution of project

Engineer has been concerned with 'reasonable' in construction practice in preference to meticulous 'literal' application of the Contract, which would need to comply with legal interpretational precedent. The need for the Engineer and the Dispute Adjudication Board (DAB) to base their views on those of a practising Engineer is not stated. Red FIDIC, Cl. 3.5, needs to be redrafted to emphasise the 'engineer-friendly' nature of the Engineer's decisions, relating always to what was reasonable at the time of Tender.

Red FIDIC, Cl. 3.1, reverses the expectation that the Engineer may be independent of the Employer, independent that is in engineering opinion, regardless of financial relationship with the Employer which should not affect the independence in opinion of the experienced, self-confident Engineer. The terms of red FIDIC, Cl. 3.1(a), appears likely to make the Engineer's function in expressing a view under red FIDIC, Cl. 15.1 (possibly leading to termination), more difficult unless, as seems improbable, this is an 'excepted clause' in relation to the Employer's approval. The problem here occurs where the Employer confronts the Engineer with legal advice, which will tend to inhibit him from expressing a contrary 'engineer's' view.

FIDIC (1987) introduced the DAB. Red FIDIC does not make clear that reference to the DAB must await the Engineer's decision on the same issue (red FIDIC, Cl. 20.1). There appears here to be a strong risk of undermining the Engineer's authority, also in prejudicing the assistance that the Engineer might be able to provide to the project where an appropriate change in design might help in solving an unexpected problem.

2.6.3 Value engineering

Possibly the most controversial feature of red FIDIC lies in its inclusion of provisions for value engineering (red FIDIC, Cl. 13.2). Value engineering, in the sense used here, relates to the sharing (often equal) of benefit between Employer and Contractor arising from a variation proposed by the Contractor. This practice has developed principally in the US as a result of the traditional lack of continuity there between design and construction. This has resulted in conservative attitudes to design (innovation seen as inviting litigation) so that Contractors are presented with opportunities for proposing major savings by taking responsibility for major redesign. The fact that the Employer may lose half of the benefit of a more appropriate design in the first place is only half the story. Fundamentally, a major change is occurring far too late for the project to be optimised, since it should have been preceded by planning, studies and detailed analysis pertinent to the new design. When value engineering is proposed, time and concern for extra cost will tend to cause additional 'preliminary' work of this nature to be skimped so the expedient may well encounter unforeseen problems, for which the Engineer is no longer the designer. Safety may also be jeopardised and there should be little comfort in supposing that the Employer's incidence to risk may be reduced by value engineering. With a strong Engineer in charge, value engineering in matters of detail merits consideration, forming, as it may, part of a Target Contract, but it will always merit attachment of a 'health warning'. Most insidiously, value engineering seems to condone weak initial design by an Engineer with inadequate experience in construction. A caution is also needed concerning calculation of the value of savings, since the variation may attract unidentified consequential costs elsewhere. Better by far to engage a competent Engineer as designer in the first place and provide the element of 'purposeful continuity'. Where there is scope for important innovation in construction, Partnering introduces the Parties to the project at a time when their input can influence the preliminary planning and hence avoids the waste and likely corner-cutting of value engineering.

2.6.4 Provisions for uncertainty

Red FIDIC contains no obligations concerning risk assessment by either Party, so there appears to be a black and white issue of responsibility, in relation to design (by the Engineer for the Employer) and for other functions by the Contractor, including changes arising from value engineering, apart from excepted matters. Red FIDIC, Cl. 4.8, might have mentioned risk control under the heading of 'Safety procedures'. Red FIDIC, Cl. 4.9, relates to QA, which is only helpful in establishing procedures for identified issues, not in the means for identification of uncertainties and hazards which may translate into risk. The only specific reference to uncertainty in red FIDIC, Cl. 4.10, is in relation to site data. What is unforeseeable? This is the wrong question. The question really is what conditions should reasonably have been included by the Contractor in his Tender Price, possibly covered by a method statement at that time, for which there seems to be no provision. The note to Cl. 4.12 encourages risk-sharing for major sub-surface work, which is a step forward. Red FIDIC misses an opportunity at this point, however, in failing to encourage interpretational reports to be provided to Tenderers. The Contractor would of course remain responsible for his own interpretation.

2.6.5 Conclusions on red FIDIC

How well does red FIDIC appear to meet the criteria for success advanced above?

1. Full competence of the Parties depends at the outset on the Employer, with the possible need to appoint a 'surrogate operator' (see Section 2.2 above). This matter, unsurprisingly, is not addressed by red FIDIC.
2. Competence of the Engineer to provide the essential link between the Employer's objectives and their achievement through construction is essential, requiring a professional capability beyond the design of a basic form of the 'product', the permanent work. The FIDIC suite makes no positive contribution to this criterion. On the contrary, unconditional encouragement to value engineering may appear to downgrade the Engineer's competence as designer, the authority of the Engineer – who is not the designer – and his familiarity with the features of the design, construction and foresight in the elimination of problems.
3. Competence of the Contractor depends on the thoroughness of the process of selection. The FIDIC suite does not address such matters.
4. Co-operation between the Parties will largely depend upon the extent to which risks have been equitably provided for. This will also affect the response to the encouragement of 'amicable settlement' following the DAB's ruling (red FIDIC, Cl. 20.5). Red FIDIC, Cl. 4.6, 'Co-operation' scarcely develops its title.
5. Uncertainty and its manifestation in hazard and risk are beyond the scope of the FIDIC suite, apart from the note in red FIDIC against Cl. 4.12 which encourages risk sharing for sub-surface works.

The FIDIC suite makes a number of positive changes within a restricted model. It carries, however, a message that the Engineer's reduced attributes no longer deserve privileges, with the accompanying duties, as a professional. The likely consequences in lack of continuity of purpose have not been foreseen. This runs counter to the trend of successful projects where the professional element of all those concerned is deliberately mobilised for the benefit of the Project. A more radical review might well accompany the suite, addressing the criteria for success and how these could be achieved through the vehicle of FIDIC procedures or otherwise.

2.7 Highways Agency Early Contractor Involvement Scheme

The Highways Agency Early Contractor Involvement (ECI) Scheme was launched in 2001 as 'Early Design-and-Build Contractor' (Highways Agency 2001). Early involvement is the paramount feature, hence a preference for the title of ECI. ECI may be classed as a direct inheritor of the Egan principles (Egan 1998) discussed in Section 2.4.3 above). Thus, the Engineer (Designer in Highways Agency (HA) parlance) and Contractor are each appointed on the basis of the expected value that they will contribute to the project, not cost. Criteria for appointment, by an active 'driving' Client, include proven quality of performance, technical and managerial capability, capacity for innovation, commitment to co-operation and access to complementary expertise. The Engineer (the term used in the sense of one whose activities are considerably wider than the functions normally expected from a designer, while not equating to those of the Engineer under the ICE Conditions – see Sections 2.2 and 2.3) is also expected to demonstrate acumen in dealings with bodies external to the project, such as planning and environmental control agencies. Early involvement of the Contractor allows specific features essential for construction to be agreed with the same agencies without subsequent hiatus. The principles of ECI may well lead to fully developed Partnering (Section 4.5).

ECI is usually expected to be associated with a Target Contract formula for payment, whereby the Contractor earns a bonus related to savings below Target, or incurs a penalty related to excess above Target. This comparison is made on the basis of recorded costs. For the successful adoption of such a formula, it is vital that adjustment to the Target is confined to specific features previously identified; ideally there should be no adjustment beyond direct compensation for external effects such as tariffs or fiscal changes. Substantial causes for uncertainty in cost should, therefore, be separated from the Target Cost into a contingency fund, drawn upon only as, and if, risk develops from uncertainty. The principle of a Target may then be focused, as intended, on the encouragement of improvements in performance, in design as well as construction. The interests of Client and Engineer are thereby directly aligned with those of the Contractor. It is to be noted that, through this form of ECI, the relationships between Client, Engineer and Contractor are grounded on professional principles.

3
Engineering and the institutions

3.1 The wider scene of engineering

Chapters 1 and 2 recount some of the developments in engineering in Britain which have contributed to a large number of institutions, each with its own constituency but with major overlaps between them. At a time when many of these represented the interests of engineers in a specific industry, such an arrangement might have seemed to make sense and it certainly had an historical explanation. The arrangement has by now been largely overtaken by events. If engineering institutions are to retain relevance to the twenty-first century, therefore, reform is overdue.

Many engineers are working increasingly beyond the identifiable disciplinary boundaries of their institutional qualification, while others remain more concerned with developing the applications of sciences, not yet formally admitted within engineering, which have no respect for differences between industrial applications. The choice is either for major reform and co-ordination between institutions or for many to wither away, probably to the loss of the profession, with their functions partially transferred to more appropriate but less industry-focused organisations.

The main failing with the major engineering institutions over recent years is that they have concentrated unduly upon the technical contributions within a traditional notion of their particular expertise, to the detriment of the widening scope and professional interests of their potential members. Risk, discovered in one field, arising from innovation or other cause, is not readily transferred to other fields whose learned society activities are focused on different institutions. Engineering and management have been allowed to fall apart, with curiously little significant overlap between protagonists of each aspect of the art.

External critics observe the rivalry between institutions, not their shared endeavours. Meanwhile, many of the interesting engineering innovations occur across engineering disciplines and beyond, spreading into elements of applied science not formally related to an institution; these developing trends are allowed to fall through the gap between institutions. As a consequence, increasing cohorts of engineers may find technical focus in journals, web sites and in proliferating conferences but with no chartered body to guard education, learned society, qualification and professional standards. The first recognisable major lost cohort was that of offshore engineering, which would have required smart anticipation by the institutions if an identifiable focus were to have been devised against the strong commercial trans-Atlantic influence in the 1950s.

Even in recent years, far from indication of inter-institution co-operation to tackle such development, there has been increased competition, at its extreme marked by gathering signs of attempted poaching. This is unhelpful in provoking mutual distrust, feeding the outer perception of an unco-ordinated and tribal set of disciplines in the profession.

Despite the technological diversity, we continue to belong to a single profession of engineer, who develop original concepts and exercise technical judgement based on the predominantly physical sciences. The range is increasing, with striking contrasts between those who work at nanometric (10^{-9} m) scales and those who continue to design and operate a world visible to the naked eye.

The combined resources of the institutions could be markedly greater than the sum of their parts; at present it is markedly less. It would be fruitless to try to set out an ideal model for the professional engineering institution(s) of the future without taking account of need to build upon what we have, enhancing the strong parts, eliminating the weak. The objective must be to perceive the needs of the profession of the future and to recognise the merits of flexibility to adapt to the changing demands of the future.

In 1977 the *Government Commission of Enquiry into the Engineering Profession* (Finniston 1978) was established to explore the future needs of industry from the profession. This was set up by the Department of Trade and Industry (DTI). At that time, the sponsoring departments for the construction industry were jointly the Department of the Environment and the Department of Transport. In consequence, it was never clear, throughout the life of the commission, whether they were addressing the engineering profession as a whole or only that part associated with manufacturing industry. The ICE nevertheless accepted

Engineering and the institutions

invitation to give evidence to the Commission. The gist of this contribution was to emphasise the need for strengthened linkages between technical engineering and management. One feature of the institution's evidence provided encouraging statistics within construction of the extent to which the upper levels of management continued to be occupied by chartered engineers, a feature of concern of the Finniston Commission. This was shown to apply, at that time, to the public sector as well as to firms of consultants and contractors. Our delegation was received by three members of the commission, one of whom soon dozed off to sleep. The report of the Commission makes no reference to our evidence and we may probably conclude that construction was in fact excluded from their field of interest. This was one of several lost opportunities for addressing the wider question of the unification of the engineering profession.

The history of the Council of Engineering Institutions (CEI) provides another example of a lost opportunity, largely on account of the policy of treating as equal the 54 constituent institutions, of which 16 possessed Charters for conferring professional qualification. The consequence was predictable, much bureaucracy and internal friction, with progress towards common purpose obstructed by the different agenda of the several factions, against not a little suspicion of *divisere et imperare*. Comparison may be made with the current attempts of the Committee of Vice Chancellors and Principals (under a new name) to present common purpose across the diverse interests of the multiple, and apparently constantly breeding, numbers and types of institutions now designated as universities!

Establishment of the Engineering Council (EC) provided a new body to supersede part of the function of the CEI, while another part, the upper reaches of the profession, was embodied in the Fellowship of Engineering, subsequently retitled as the Royal Academy of Engineering. The EC has made contributions to the projection of engineering to the world beyond and to the raising of the academic standards of entry as chartered engineers.

In 2002, the EC split its functions of regulating the engineering profession (now designated EC^{UK} to emphasise that this role is confined to Britain) from the wider role of championing science, engineering and technology through the Engineering and Technology Board (ETB). The primary targets for the ETB are to attract more talent into engineering and to enhance the contribution of engineers, working in association with others with complementary functions to the same objectives. The work of the ETB receives a matching grant from Government,

Civil engineering in context

with 40% of the engineers' registration fees to EC^{UK} and 60% to the ETB. This is a move which deserves full support from engineers through their institutions, but needs rapidly to become more visible. It also requires a counterpart of institutional co-operation.

A visionary report on behalf of the Royal Academy of Engineering and the EC (RAE 2000) explores many areas in which the profession could help to achieve the aims of the Hawley Review Group 'to review the contribution the Council (i.e. the EC) should make to add value to the engineering community, to the benefit of the UK economy'. Emphasis is given to the need to break down divisions in commerce and elsewhere between the creative, organisational, managerial and business skills inherent in the role of the successful engineer.

To many, however, the initiatives of the EC through SARTOR (Routes to Qualification of Professional and Technical Engineers) to control entry are seen as too much dominated by academic degrees, as opposed to demonstrable, and demonstrated, acquired capability. In 1999, reflecting the lowering standards of entry to engineering courses at their lower end, the EC adopted rules of mean A-level scores at entry to degree courses to determine whether they should qualify to lead to Chartered or Technician (renamed Incorporated) Engineer. If this were a fair basis of sub-division, then it must appear that prior to this ruling a considerable proportion of Chartered Engineers should not have been so designated. Or does it? The answer must depend upon the standard of scrutiny observed by the individual institutions at the time of undertaking the professional interview and examination for admission of a particular brand of chartered engineer. The reaction has been predictable, with a reduction of enrolment to the weaker schools of engineering on the one hand with the appreciation that these no longer provide a direct route for the chartered engineer, while on the other hand institutions perceive a threat to the size of their membership.

We have currently the bizarre situation of a shortage of middle-grade engineers such that recruitment from overseas is seen as a short-term measure, but coupled with continuing inadequacy in our recruitment of students to civil engineering. Several factors have contributed to this situation:

- The high relative pay of other professions, e.g. law, accountancy, with less rigorous entry qualifications. By any objective measure, those in the commercial world overpay themselves. Salary levels in civil engineering, nevertheless, represent a casualty of least cost

prevailing over value-for-money and lack of perception by management to identify and encourage the most promising engineers.
- Lack of confidence in long-term investment in construction and the environment.
- Inadequate enthusiasm aroused by secondary school maths and science teaching by those with little interest and negligible understanding of applications.
- Instances of poor project management coupled with the British exuberance of the media in providing news about disasters unbalanced by interest in success.
- The failings of we, the professional engineers, to project the positive features of our chosen profession.
- While issues of sustainability, with environmental, economic and social dimensions, acquire increasing interest, there is an absence of public awareness that effective delivery of political aspirations depends predominantly on engineering. Too many learn about the problems; too few about the solutions and how they may be achieved.
- Destruction of the routes into engineering through technical colleges and Higher National Diploma (HND), for example, which attracted, through obviously relevant instruction and training, those who are deterred by courses attempting abstract rigour for the unscholarly, but find satisfaction in the practical achievements of engineering.

An obvious self-inflicted injury has been the centralised uniformitarianism imposed on higher education, in the false belief that this provides equality of opportunity. The best of the polytechnics provided HND courses in parallel with degree courses with interchanges related to capability, especially useful in identifying the late developer, held back by family, school or peer-group background. The vocationally-oriented HND would be more satisfying and useful to the student interested in the practical features of engineering but with limited grasp of theory. The more analytical course, appropriate for Chartered Engineers, is a positive disincentive when taught as a matter of rote to more practical-minded engineers. Such people would find greater satisfaction in a more practical course, leading to interesting careers, of greater value to the industry. The possessor of a low-grade academic degree may continue throughout his career to be unable to relate his academic education to his occupation, never quite understanding the principles of what he has been taught. The engineering profession draws strength in creativity from the variety of the talents of the members; the minimum shared capability must be based on the ability fully to communicate.

Whether or not engineers are underpaid depends on the criteria for judging their performance. The dependence on good engineers for the success of the industry and for the public services will not be questioned. It is also a fact that much of the disappointment in performance is attributable to conditions in which a project has been put together by others who have inadequate understanding of the marks of success. Of course, the engineers should be in positions of greater influence to help to avoid such occurrences. The engineers are blamed for failure but success is too often attributed to others. This again is partly the fault of the reticent engineer. The engineering contribution to the millennium dome was outstanding; so has been the contribution of engineers to the protection of the leaning tower of Pisa – although the media, including the BBC, gave the credit to architects and archaeologists. Who noticed that the splendidly eye-catching eye-lid engineer-led Gateshead Millennium Bridge across the Tyne received a special award by the Royal Institute of British Architects (RIBA), with Marcus Binney of *The Times* attributing the project entirely to the architect, without mention of Gifford, the engineer? (Persistent letters to *The Times* to expose such solecisms are neither acknowledged nor published.)

Attention to the causes of failure to attract more talent into engineering would bring with it solutions to many problems. These would include: increased esteem for the engineer; positive emphasis on engineering as a career; greater attention to the different routes appropriate for a balanced workforce of engineers attuned to the challenges and performance of their profession. It is an unacceptable and expensive admission of failure if we continue to depend on importing a substantial fraction of engineers educated and trained elsewhere. As a practical issue, unless we improve our own practices, such a temporary resort only adds to the long-term problem. The very notion of a doctrinaire approach to equality of opportunity leads to the denial of opportunity where different motivation and capabilities are ignored by the doctrinaire route of political correctness.

Largely conditioned by the model of the commercial, more specifically the financial, world of personal ambition undiluted by mutual loyalty, or even respect, between employer and employed, the notion has gained support that constant mobility, associated with the acceptance of risk of instant dismissal, is a mark of success. Evidence of devotion to duty in the commercial world is increasingly related to working excessively long hours, at the expense of a balanced life-style and to the obvious detriment of the quality of work achieved, with diminished time for thought or innovation. The notion of loyalty is even expressed pejoratively as an anachronistic recollection of a bygone age, to be replaced by more competitiveness. For a profession, all this is nonsensical. In the early days of a career, variety of experience and exposure are desirable, which may entail a change of employer. Generally, a shared interest in the career of the individual, his inter-personal responsibilities and the future prosperity of the organisation will lead to a relationship of support and mutual trust combined with a sense of security. The shared ethic and sheer addition to efficiency resulting from such stability should be reflected in the 'value added', with benefit both to organisation and to employee. Of course, investment for the future is not a liability for the improvident, predatory examples from the commercial world. The successful professional engineering organisation finds a middle way, which combines the virtues of a stable work force with an element of external recruitment to overcome excessive dependence on self-sufficiency in talent and creativity, the not-invented-here syndrome.

One of the most striking developments in civil engineering during the last 100 years has been the change – where it has been permitted to occur – from commercial to professional relationships between those organisations engaged in projects of construction and operation of facilities. At one time, as discussed in Chapter 4, the professional was the consulting engineer, recognisably an individual who employed engineers and others. Other relationships, based on terms of contract drawn up by the consultant, were of a strictly commercial nature. The work to be undertaken was specified – up to a point – with its means for execution largely dependent on artisan skills. For the major successful project today, an essential element is that of the mobilisation of a variety of professional capabilities from many engineers, in whatever capacity they may be involved. They are contributing ideas towards success, not just delivering specified components. The consequences of such change are discussed in succeeding chapters.

3.2 Unity of the engineering profession

For many years the author has been a strong proponent for moves towards unification across the profession of engineering, with the objectives of:

- Co-ordination of activities of individual institutions in the interests of their membership, technological advance and economy.

- Encouragement of those whose careers take them across the traditional boundaries between institutions.
- Increased regard for the views of the engineering profession on questions of policy.
- Development of inter-institutional policies for new areas of engineering.
- Rationalisation of qualifications and learned society activity across the full range of engineering.

A fresh opportunity for such a development occurred – or seemed to occur – when an initiative towards unification was prompted by Dr J. Fairclough, as Chief Scientific Advisor to the Cabinet. Evidence was widely sought. My own contribution was based on the interests of individual engineers, relating to their institutional requirements, as opposed to the top-down approach which seemed to be developing. My proposal received no mention in the subsequent report.

Essentially, my proposal started from the recognition that the great technical diversity at the start of engineering careers converges as the duties of senior engineers working in different disciplines include increasing degrees of management through the operation of systems. In a discussion document, the term 'colleges' had been used to describe the grouping of institutions within cognate disciplines. I picked up this concept but applied it in a somewhat different manner.

My proposal advocated a model built around the four major institutions: civils, mechanicals, electricals, chemicals. Each group would establish particularly strong associations with specialist institutions in their area of interest, starting with a rationalisation of learned society interests. Several specialist institutions retain strong relationships with particular industries and these are, therefore, able, at least in principle, to provide learned society support of specific interest to a wide range of Chartered Engineer (CEng) and Incorporated Engineer (IEng) and others in these industries. In a similar way, specialised learned society interests are served by several societies based on the ICE.

The recognition of increasing trends towards new branches of engineering, either resulting from the increasing breadth of engineering or with the development of applied science to recognised technology, requires a new machinery which crosses the disciplinary groupings described above. It is here that I adapted the notion of colleges that had been introduced to this. Wherever the operational field crosses the disciplinary scope of one institution to another, whether or not the field remained within the disciplinary grouping, a 'college' would be established to cover the special interests. The college would develop schemes for qualification, and to consider how best to look after other special interests such as the learned society. Each college would be allocated for administrative purposes to a named major institution. Additionally, there would be increasing use of engineers from other groupings to take part in the college and in the appropriate professional assessment of candidates. The choice of route for qualification would depend on the general leanings of the candidate and expectations of direction of career development.

The general direction of engineering education must recognise the increasing areas of common ground between the several disciplines. For some, this may entail, for CEng, introduction to risk and to early working on projects, which bring the disciplines together. Elsewhere, for a few universities attracting those with special analytical strengths, the approach may depend on a general engineering science and design course, with later specialisation. Others may continue to teach more specialised courses from the outset for CEng and, in a more predominantly vocational manner, for IEng. Such an arrangement should help to fulfil the requirements of industry, motivate and satisfy the interests and career needs of students while increasing the attraction to engineering courses.

A unified profession of engineering would aim to rationalise rules of conduct and of qualification between institutions and would provide a vehicle for advice to Government, ensuring that different emphases from different viewpoints were effectively combined, providing a single view from the profession, with indications of uncertainties. Thus, the excuse for disregarding the advice of engineers on the grounds of contradiction would be removed. This function would be co-ordinated with the Royal Academy of Engineering and the ETB.

A trend towards unity of the engineering profession, in recognition of the increasingly uncertain frontiers between and beyond current institutions, appears inevitable and should be planned in preference to awaiting the collapse of the present system. If the institutions fail to take the initiative, they may well find themselves outflanked by calls for a statutory body to control the profession, a call of the nature which largely motivated the Finniston Commission.

3.3 Evolution of the Institution of Civil Engineers
An excellent history by Garth Watson of the ICE (Watson 1982, 1988) exists and I do not intend to cover the same ground. Those of us with

long enough memories will picture the Institution of old, run by a council as senate, headed by a President in his retirement year or beyond and with the setting presenting all the outward appearance of a gentleman's club. Sir Harold Harding recalls that only the most intrepid engineer (of whom he was one as a young man and survived as such into old age) outside the age group of the senate would dare to express a view in a technical debate. Traditional practices were changing slowly and incremental progress depended on the collective received opinion based upon recollections of practical experience. The horsehair stuffed settees of the lecture theatre would have made a perfect setting to reproduce the Royal Geographical Society for the report by Burton and Speke on the discovery of the source of the Nile in 1859.

Institutional changes in custom and operation were inevitable. Engineering finesse depended increasingly on the applications of science, with complexity matched by facility in analysis aided by electronic development. There was a period of relatively rapid technological change through, say, the 1970s, reflected, for example, by succeeding editions of *The Finite Element Method* (Zienkiewicz 1967, 1971, etc.) combined with a new generation of engineers whose potential was stimulated at university and beyond to match the rate of technical advances of the computer.

A by-product of this rate of change was undoubtedly a contributory factor to problems in construction, for example of box girder bridge failures. Here, the traditional bridge designers experienced problems in communication with the young analysts who were armed with powerful tools but who lacked experience in addressing the vital questions and, hence, wielded their tools too crudely. This problem becomes particularly acute in the intermediate stages of construction, which impose duties upon the structure different from, and in certain respects more onerous than, those of its operational demands, requiring moreover answers in real time. This was a difficult period of transition in design for all concerned, while computer-based calculation by the few proceeded in the company of slide-rule engineering, combined with manual calculation and draughting, by the many. Increasing complexities of construction at site passed through a comparable period of change in nature rather than degree.

The institution responded by a great increase in its learned society activities through an increasing range of special interest groups, reflecting the growing diversity of the occupations of the civil engineer and the development of new and specialised techniques of design and construction. Naturally, as a consequence, debates in the institution commanded a wider range of age and skills. Interests also developed in the wider responsibilities of the engineer – in society, the environment and in appropriate technology – as reflected by my own Presidential Address (Muir Wood 1978). There have been several indicators of the search for new perspectives arising from such changes in direction and scope. Under the Presidency of Dr E. C. Hambly, for example, *The Institution as a Learning Society* (never published, as a result of the sudden death of the President), considered the two-way learning process between the institution and its membership. The green credentials of the institution were displayed under the Presidency of Mr R. N. Sainsbury (ICE 1996a). Under Professor T. M. Ridley, the future direction of the institution and the profession were reflected in *Whither Civil Engineering* (ICE 1996b). These are scarcely the signs of an institution deserving the rebuke from its Secretary in the Annual Review for 2000: 'We are rapidly changing from dwelling on our historical background to preparing for the future in which we will be working' as if this were a new discovery for the 21st century. Greater familiarity with the institution's history would have indicated that the comment was unjust to those who, throughout its history, have endeavoured to look towards the needs of the future, sometimes far ahead. The temptation to mark the new millennium with suggestion of revolutionary change was evidently irresistible.

At a time when, to the public at least, initiatives for change were coming principally from outside the institution, e.g. the Latham and Egan Reports, the profession could well have appeared to be on the defensive. Appreciation of the widening role of the civil engineer was only slowly becoming understood by engineers, whose education and training was largely confined to technical elements. Moreover, morale had been sapped by the consequences of fragmentation of project management of the 1980s. From such features, accounts of projects described litigation not success, recruitment of talent was diminished, career opportunities for construction appearing to wane against the brasher self-confidence of the law and finance.

The corner has now been turned. The civil engineer accepts the multiple dimensions of his tasks (if not the bureaucracy that attends the working of some of the external agencies). Major projects based on professional principles provide benchmarks and career satisfaction. The institution once more takes the initiative in leading change for the new contexts of civil engineering.

3.4 ICE Policy on reports and publications

From time to time I have criticised the institution for inadequate coherence between reports on specific aspects of civil engineering, particularly in the intimate relationship between technical matters and management. Each sponsored report, in consequence, may tend to be self-contained or to stray into areas lacking a coherent overall philosophical strategy. The problem seems to be compounded by the absence of a body, internal to the institution, with a continuous function of development and presentation of an 'institution policy for the profession'. The Executive Committee cannot be charged with such a function since it is too much concerned with administrative business, especially at a time of constant change. It could well be that an institution 'think-tank', drawing upon accumulated wisdom and experience could fulfil such a function; its members would probably recently have retired and would have, at least in principle, time for recollection and reflection – but they must be forward thinkers, not fossils. This body would contribute to the terms of reference of commissioned reports and, as these proceeded, to guide their authors so that the report may contribute positively to coherent overall institution policy. This is not a question of thought control but, more often, of reports wandering beyond the intended scope or authority of the authors, without anchorage to a statement of good practice. The think-tank would be an advisory body, possibly known as a Senate. Its own framework documents would, after discussion and acceptance by the President and Council, provide guidance on professional policy until such time as revision were deemed advisable. The constant aim would be towards developing understanding of the holistic criteria for good practice. The American Society of Civil Engineers (ASCE) for several years published, among its proceedings, the *Journal of Professional Issues in Engineering Education and Practice*, which contained precisely the type of paper that could encourage reflection on the wider issues of our profession. Sadly, this journal no longer exists and there is no other journal addressing this broad environment of practising engineers. There was never greater need for the expression of principles of such nature, extending well beyond the syllabus of instruction of a present-day civil engineer.

Traditionally, the ICE encouraged the presentation, and subsequent publication with discussion, of accounts of the execution of major projects and of research of practical application. Reports have also been commissioned on specific subjects. Many of these Papers remain sources of great interest to the biographer, the historian and, particularly, the engineer for accounts of problems, foreseen and unforeseen, with the means for their solution. The major innovations in civil engineering may be derived from these sources. For example, two classics are to be found in Works Construction paper No. 13 on 'Compressed Air Foundations' (8 March 1949) and a subsequent debate on the relative merits of methods of sinking bridge foundations of 12 December 1950.

These Papers had several virtues:

- They were refereed.
- The most important were open to verbal discussion (and all to written comment) which was published with the author's reply.
- The scope often embraced issues of interest to the researcher as well as the practitioner.

As computer methods advanced, so were the Proceedings increasingly invaded by Papers relating to the manipulation of matrices and similar mathematical features of numerical solutions. Mistakenly, in my view, research papers were then segregated in 1974 from 'works' papers, thereby reducing the objective of encouraging practical applications of research. As might have been expected, published research became increasingly esoteric, published practice increasingly pragmatical. This division was reversed in 1992, when a new format set a standard, more readable style, with the Proceedings split into fields of application, a 'horizontal' as opposed to 'vertical' cleavage. There have also been special issues blending design with construction on major projects, such as the Channel Tunnel and the Øresund Crossing, in which British engineers played a major part.

High presentation standards of accounts of projects and of techniques undoubtedly encourage the attention of readers. The essential feature of any account is to identify issues that may concern comparable situations elswhere, the particular problems, their solutions, the relationships between the participants and so forth. The responsibilities of the editors of learned society journals are considerable: to ensure adequate depth of analysis; readability; and reasonable objectivity. Where a project is well planned, executed and administered, all parties are usually ready to contribute to an account which will explain the particular features of success, with accounts of problems and attributions to other participants. It may generally be safe to conclude that major projects without adequate record in ICE publications are those that have disappointed expectations of their promoters who may, therefore, have discouraged publication of professional, as opposed to popular, accounts of their projects.

It may be suggested that the institution no longer bears a duty to publish Papers of the most important works and research. Any subject presented to a search engine on the Web will provoke quantities of information – so why seek more? The Web is subject to no winnowing of the reliable from the spurious so it is good for quantity, but quality is elusive. There are, of course, increasing numbers of specialist journals which address their own fields and these have filled the gaps that would otherwise exist. These articles are not, however, rigorously refereed – if at all – and there is too often a detectable commercial slant that obscures objectivity; to impress takes precedence over the need to inform. Of course, there are honourable exceptions.

There is an accompanying lack of reliable reference books. Those that are commissioned and published are usually of good quality, providing balanced coverage of the claimed ground. Too many, however, with the title of a reference book are in fact records of conference proceedings. The defects of these are two-fold and my concern here is with accounts of value to practising engineers, not proceedings of research papers, which are of a generally higher standard. The defects of volumes of 'practical' conference papers are these:

- Ineffectual refereeing and a high commercial objective. As a personal experience, I have often found that my objections to a paper of low quality have been overruled not, I believe, that my criticisms have been found irrelevant but because the conference organisers do not wish to deter participation by the author to aid viability of the conference.
- No attempt is made to cover fully the area of the conference so that wide gaps exist in the presentation of the subject area. State-of-the-art Papers by those with wide experience and sound judgement are increasingly rare.

There are occasional papers of high merit in conference proceedings but generally the seeker after knowledge will be disappointed. There is rarely an index, rarer yet at the present day a summary that selects aspects of merit from contributions. The enquiring reader depends on the identification of a reliable name among the authors or on a discerning librarian.

Through this period of variable standards of published records, the institution journal *Géotechnique* maintains an enviable international reputation, with thorough refereeing undertaken within a commendably short time frame. The contributions by engineers of the editorial board and institution staff merit emulation. A new problem here is the submission of too many Papers of low merit – which need to occupy excessive time in review and rejection – probably as a result of the experience of their authors of undiscerning acceptance for Conference proceedings. Possibly, two-stage refereeing might ease the problem, a rapid sifting by a single experienced reader to separate the, at least potential, wheat from the chaff.

From the examples described above, the case is surely made for a stronger, continuous but flexible, policy of the institution in relation to its learned society role, especially concerning publications. This need is closely related to that of a coherent professional policy, for which a body such as the 'Senate', described at the opening of this section, could contribute to a solution.

3.5 Qualifications for membership

What are to be the future criteria for membership? From an origin in which the civil engineer might be defined as one engaged in a multitude of technological activities outside military engineering, the definition of civil engineer became more confined to construction, with the birth of other professional institutions. As discussed in Section 3.2, many engineers now work across the traditional boundaries of the major institutions. Many others engage in more specialised activities which, as new developments, escaped definition as civil engineering. Leaving aside the definitions of the criteria for CEng or IEng, there appears to be scope for broadening the membership. There needs to be strict rules for such development, for efficient working and to retain esteem for standards. The common denominator for the capabilities of a Chartered Civil Engineer has already shifted markedly with time. The specialist complements the generalist with limited areas of common ground.

A 'descriptive engineer', i.e. one who relies upon no more than accumulation of undigested data, is a contradiction in terms, an oxymoron. There must be an adequate measure of ability for science-based analysis. This capability must reliably inform the engineer upon conditions in which he is competent to rely upon his own judgement or where, alternatively, he needs to recruit or consult other resources. A major consequence of good education and training is to know what one does not know. Figure 5.3 illustrates the levels of understanding to enable members of a project team from different backgrounds and fields of specialisation to work effectively together. Such a diagram might well provide a model for the acceptance for membership of

Civil engineering in context

those who are engaged in activities related to construction but not along traditional lines. The criteria might include:

- *expertise* in the particular field,
- *competence* to work with 'traditional' civil engineers, understanding the basis for their functions, if not intimate acquaintance with their technologies,
- *awareness* of the roles of others in the evolution of a construction project,
- demonstrated ability to conceptualise from experience, i.e. to understand where parallel examples of application may occur.

Differential degrees of capabilities in such respects would be expected from candidates, depending upon their roles. Each would be expected to demonstrate ability to apply a mathematically-based scientific approach to his own activities. Rules of this nature would generally ease the 'conversion' from a first non-engineering degree. Proven capability would remain the primary qualification; this formula is not advocacy for easy entry by a side door to the profession, but rather a strengthening of the profession of civil engineer.

A broadening of membership in this way would bring indirect benefits in gaining greater access to formers of policy and opinion. Too often the civil engineer appears too late upon the scene to correct decisions in basic planning, made by economists, for example, who have not understood their full consequence. Changes would obviously need to be considered as part of a system, to be explored before implementation to ensure mutual support and compatibility.

3.6 An historical perspective for the profession

I recall the opening sentence of my own Presidential Address in 1977 when I remarked that 'The sheer continuity of our long-established profession, with its origins traceable to prehistory, can too readily blind us to the need, from time to time, to reflect upon the changes taking place in society and to consider the extent to which the attitudes and endeavours of our predecessors remain relevant to the present and to the future'. I then considered responsibilities of the engineer to society and for the environment, drawing attention to John Stuart Mill's notion of the mid-nineteenth century for the 'sustainable society', before the term was popularised through Mrs Gro Harlem Brundtland of United Nations Educational and Scientific Committee (UNESCO) in the 1980s. I make no claim for egregious concern for reform; many of my predecessors have done so in their own time. By changes in the by-laws of 2000, the membership gives Council greater powers of autonomy without reference back to the members; the condition to be expected for this indulgence is that the powers be exercised wisely and with valid explanation to the membership for the reasons for projected change. The greater the degree of communication between institution and membership, the less the likelihood of any sense of shock when a significant change is proposed. As the President travels around the regions, and the world, valuable opportunities are provided for discussion and familiarity with views of the members which may not be apparent at Westminster.

The President of the time is in a privileged position, in so far as other commitments may allow, to reflect upon the state of the profession with a wide audience and a published record. The Council, Officers and Staff of the institution are constantly concerned with the interests of the Profession, to ensure that conditions of membership remain relevant to these activities, to encourage participation in the affairs of the institution and to widen the appeal to other potential members who have hitherto appeared to be discouraged from joining. The President is expected to lead by example and, unless clearly making an exception, is speaking for the profession. If he feels the urge to express personal views that vary from those of the institution, he must ensure that this fact is stated explicitly. It is particularly damaging to the standing of the profession if he is apparently speaking *ex cathedra*, but expressing his own views, which run counter to institution policy, especially when his statements may appear to correspond to his own commercial interests. The President is representing the profession as competent, objective and trustworthy.

The functions of the institution have grown from those of a club, largely conducted by amateurs, to a complex organisation, its activities highly geared together, with many facets of professional and technical interest led by Vice-Presidents, co-ordinated by an Executive Committee. The associated actions to provide support for the members' activities, budgetary and administrative functions are overseen by the Director-General who occupies a pivotal position in planning for the institution of the future. I have known personally more than 50 Presidents, several of whom have made outstanding contributions, working with their contemporary colleagues and officers, to the evolutionary style of reform. I remain confident that their work will continue effectively to be built upon.

4

Engineering, management and the law

4.1 Introduction

Perhaps I should open this chapter by insisting that I continue to enjoy personal friendships within the legal profession. My complaint is that as a profession they have contributed more to the lucrative exploitation of the defects of contract procurement, project relationships and project management than to their elimination. I should here except the positive activities of the Centre for Construction Law and Management (CCLM) based on King's College, London, which performs valuable services by bringing together engineers and lawyers in courses and conferences. Its example is being followed elsewhere around the world. A mutual understanding is at least the first stage towards a sounder recognition of the desirable roles of engineers and lawyers respectively in the interests, primarily, of the Clients, those served by the construction industry. It deserves note that even this new breed of construction lawyer is being left behind by the rapid progress in professional relationships described in Chapter 2.

During a single career of a little more than half a century in civil engineering, the influence of lawyers to civil engineers in their professional role and to the construction industry has changed not just in degree but in principle. My object is to discuss such changes, provide some advice on the trends and suggest criteria for measuring whether and where their influence has been for the good. The law has much affected both contract procedures and project management. The two are sufficiently intertwined not to attempt their comprehensive disentanglement.

Rather, my purpose in succeeding paragraphs is to describe past practices in contract administration adequately to emphasise positive features which merit preservation and, indeed, development for a healthy construction industry for the future.

4.2 Principles of good and bad practice

During the immediate post-Second World War period, in Britain the civil engineer as Consultant expected to be the first to be approached outside a Client authority about a prospective project. The strategy would be discussed, with many aspects of financial needs, resources, land needs and estimates of income (where appropriate) prepared jointly. Lawyers were at hand but appeared to be consulted on specifically legal issues, not on construction policy. Projects were undertaken on the basis of contract terms prepared by the engineers (the ICE Conditions and later the overseas FIDIC version of similar style – see Chapter 2). Differences were settled by the Engineer in a quasi-judicial capacity conferred by the Contract. If, occasionally, matters fell into dispute, the arbitral function would be conducted by an engineer of long experience in the field of work, with evidence presented directly by the parties involved, with possible inclusion of outside expertise as needed.

Those who did not participate in this world, or only observed it in decline when the procedures were becoming abused and misused, describe the practices as confrontational or in lacking adequate co-operation between the engineer as designer and the engineer as constructor. It was, on the contrary, an excellent system when operated with proper professional acumen and objectivity. This professional attitude was expected from all parties concerned. What yet remains imperative, since no general formulation has achieved this objective, is to preserve the virtues of the system, not necessarily the system itself since the world moves on, so first we need to define what these were and remain. Without premature disclosure of too much of the plot, it may be averred that in many ways these correspond to the features of successful examples of project management, which are currently building on the virtues of the ICE Conditions when operated with professional skill and good intent.

Onora O'Neill began her BBC Reith Lecture 2002 with this account:

Confucius told his disciple Tsze-kung that three things are needed for government: weapons, food and trust. If a ruler can't hold on to all three, he should give up the weapons first and the food next. Trust should be guarded to the end.

I allude more than once in this book to the essential feature of trust, needing to be earned and rigorously guarded by the engineer. I leave to the lawyers their own concern with this virtue.

4.3 Decline of the ICE Conditions

The loss of confidence in the ICE Conditions, mentioned in Section 2.3, was essentially a loss of trust in their operators. As already outlined in Chapter 2, this arose from several causes:

1. The jealousy of the Client in relation to the independence of the Engineer. There were suspicions that additional costs were arising from mismanaging the project and, particularly, in deficiencies in design, which might contribute to increases in cost, being laid to the Client's charge, without recognition that the Engineer as Designer might be expected to bear a measure of liability.
2. In certain circumstances, on the other hand, the Contractor might have grounds for complaining that a weak Engineer, in giving 'independent' rulings, was allowing his own views to be unduly influenced by those of his Client. As a variation of this circumstance, there were known occasions in which an Engineer, having given assurance of a project being undertaken within a certain sum, rejected valid claims, which would have caused this figure to be exceeded; this, of course, was thoroughly unprofessional. The problem was exacerbated by major Clients who revised the ICE Conditions deliberately to curtail the independence of the Engineer, thus destroying the intended balance of the form of Contract. Perilous was the position of the Engineer whose powers were curtailed without revelation of this fact in the Contractor's Conditions of Contract.
3. High levels of competence were expected of all Parties. The traditional mould comprised a few experienced Clients for major works, well-established firms of consulting engineers and contractors, who were familiar each to the other. As this community was invaded, with the rapid increase in work, by less experienced newcomers to the field, so did the mould break and the professional element find increasing difficulties in withstanding the new commercial pressures while maintaining quality of service.
4. A scale of professional fees, never mandatory, helped to avoid competition on price. Once this safeguard, seen by some as an unjustifiable professional privilege, was breached, the way was open for competition on price and general collapse of the system, only exacerbating the trends of 3 above.

Each of these issues reflects on professional competence and integrity, ultimately on trust. The competent innovative Engineer knew that he was giving good value to his Client. If he were negligent he would expect to have to pay for it. If, however, there were a situation in which he had taken a known risk in the expected interest of his Client (and, of course, this was helped by the degree of mutual confidence) he would explain the situation and look for sympathetic treatment. The community was sufficiently small for peccadillos or more serious infringements to become known in the community, not yet protected behind parapets of commercial confidentiality. The greatest loss to the professional engineer was that of professional integrity. The system, under attack and essentially holed by competition on price between consulting engineers, largely collapsed in the 1980s when extreme market-driven politics of the time directed the structure and priorities of the Public Sector Client and, most particularly, the newly privatised Statutory Authorities. Thus arrived the practice of turning to lawyers and accountants, in preference to engineers, for procurement strategy in general, including construction in particular. There are many facets to this development, possibly the most important concerning risk. Bluntly, where uncertainty is converted into exposure to risk by deliberate or uncomprehending action, a high price has to be paid for this adventure.

4.4 Application of principles of good practice

The skill of avoiding the progression: uncertainty → hazard → risk is, or should be, a valuable attribute of the engineer. Perhaps a few definitions for the context should be introduced here:

> *Uncertainty*: the extent to which features that may affect the scheme of construction remain to be defined or determined, if they are not to contribute to hazard.
> *Hazard* (from BS 4778): a situation which could occur during the lifetime of a product, system or plant that has the potential for human injury, damage to property, damage to the environment or economic loss.
> *Risk* (from BS 4778): a combination of the probability, or frequency, of occurrence of a defined hazard and the magnitude of the consequences of the occurrence.

The lawyer considers it a virtue to avoid exposing his Client to risk, in the absence of understanding the cost of so doing or the means for its mitigation or control. The engineer recognises that this form of risk aversion will often incur high and avoidable cost and may, in fact, contribute to the probability and the magnitude of the risk. It will certainly discourage desirable innovation. These different approaches between lawyer and engineer entail fundamentally different attitudes to the relationships between the Parties to a Project, and to the Contract around which these may be established and administered. Each will recognise the need for certainty in interpretation of contract documents but this is near the limit of their common purpose. If project requirements are defined while major uncertainties affecting the execution remain unidentified in nature or magnitude, the project is launched into the unknown. The Client is, in effect, being advised to haul up the drawbridge for a long siege. I have yet to encounter a successful major project resulting from this approach. The hazards and consequent risks, uncertain in nature or degree, are passed to the Contractor – who will doubtless attempt to transfer these to others – and the stage is set for another example of the unwished-for adversarial notoriety of the construction industry.

The fundamental defect of the commercial approach to the control of engineering works is that of deliberately excluding the professional dimension and hence, incidentally, much of the benefit of the engineer's capability for understanding and controlling the system. The project is seen as a simple transaction whereby the worse the bargain for the Contractor, the better the bargain for the Employer, a simple zero-sum concept of game theory. The element of uncertainty is disregarded and the project administrator, a hands-off operator, has neither the ability nor power to remedy the faults of rigid demarcation and fragmented operations. The consequences are predictable and many examples are by now historical events, for example the Jubilee Line Extension of London Underground. Here the glue of the input from engineering design was virtually eliminated. Site investigation was one competitive transaction, construction another, design being reduced to the input of a 'design contractor'. This must be near the nadir in absurdity and the price in overruns was excessive, largely in direct consequence.

The loading of the cost of uncertainty upon the Contractor has two main consequences. First, that it will lead to unnecessarily high costs to the Client, who will have to bear the maximum incidence of each risk, however it may or may not be manifested in fact. Second, that such a practice tends to blight the relationships between participants; the economic way of dealing with an uncertainty may entail a revision to the Client's planning or design, but this is not to be expected by a Contractor to whom excessive risk has been attached. Several attempts have been made to illustrate the sharing of risk between Client and Contractor for different types of contract. Figure 4.1, after Flanagan

Engineering, management and the law

Types of contract

Design and manage
Design and build, turnkey, package
Lump sum, fixed price
Lump sum, fluctuating price
Cost + fixed fee + target price
Schedule of rates, remeasured
Management fee + fixed price works
Management fee + cost + works
Management fee + guaranteed maximum price
Construction management

Who carries the risk?
Client | Contractor

Fig. 4.1 Who carries the risk? (After Flanagan and Norman, 1993)

and Norman (1993), is an example; but this may mislead if the reader interprets the vertical chain line as representing a constant normalised contract value. Furthermore, there is an implication that the overall magnitude of risk remains the same for each species of contract. This is not so, since co-operative working between Engineer and Contractor, to overcome an unexpected phenomenon by an economic solution, will be prevented where the provision for financing the risk is placed excessively on the Contractor. Figure 4.2 sets out the data for selected forms

Minimum cost
Risk carried by Contractor
Paid by Employer
Risk carried by Employer

Relationship

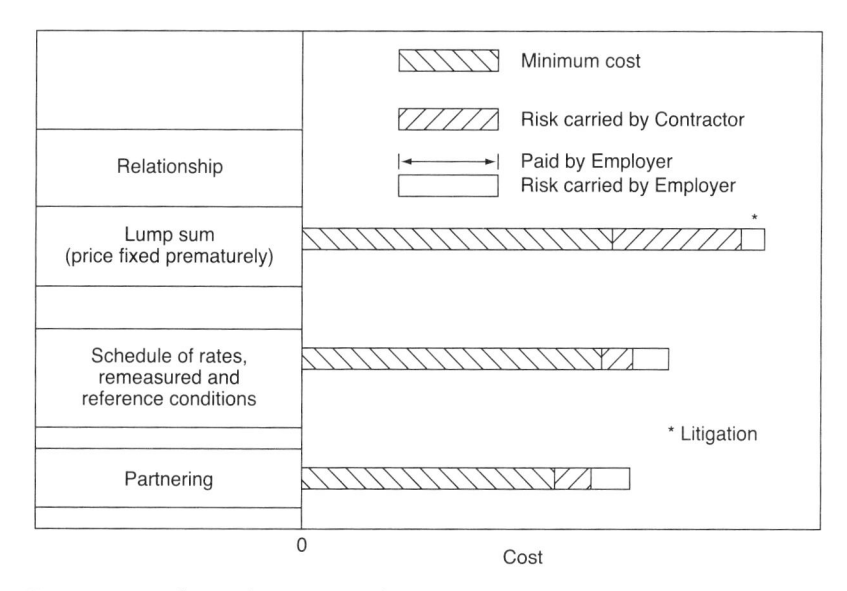

Lump sum (price fixed prematurely)

Schedule of rates, remeasured and reference conditions

* Litigation

Partnering

0   Cost

Fig. 4.2 How the employer carries the cost

65

of Contract from Fig. 4.1 in a manner better to establish the exposure of the Client – who, of course, has to pay for the risk, whoever the primary responsibility is carried by.

The New Engineering Contract, later the NEC Engineering and Construction Contract discussed in Chapter 2, contains much good sense in eliminating the undergrowth which becomes attached to any document in the hands of lawyers. The ICE Conditions, as it developed from edition to edition, tended to obscure the objectives as legal precision overreacted to engineering simplicity. To the lawyer, there seemed to be incomprehension that the increasing complexities of construction required greater protection to the Contractor from the costs of variation. To the Engineer, interpretation based upon meanings generally understood between the Parties is preferred to a micro-specific legal construction (and curiously this seems to be rather nearer to the construction of French law). The NEC provides a framework for possible continuity through a part of a project and for parts to be linked appropriately together, regardless of the engineering disciplines associated with each. The numerous options require good understanding if they are not to be misapplied. It is valueless to proclaim the virtues of collaborative working between employer, designer, contractor and project manager (opportunities which exist in many forms of contract operated with enlightenment) unless the right option is adopted. As the team is assembled for the project, so is the ethos for co-operation (a preferable term for collaboration which has a ring of compulsion) practised and developed. This cannot occur if the first player, the Client, is already on the defensive. The Designer as controller, on the other hand, is able to foresee potential departures from expectation and head off the consequences before they occur, thereby achieving a positive gain – apart from the co-operative attitudes that such anticipation engenders.

4.5 Getting the best out of Contract Conditions

Principles of teamwork, of transparency in sharing information about a project, of the treatment of information and interpretation of relevant information about the ground, are increasingly accepted for major engineering projects. What is less generally understood is that these same principles may often be achieved through reinventing the functions of the Engineer under the ICE Conditions. The Engineer, appointed for the work directly, as a result of interview or by establishing availability of resources and experience relevant to the project, was not at the outset subjected to competition by price from those with lesser moral fibre and experience. In consequence, it was possible to give objective advice to the Client for his policy in relation to the project, whether or not this might affect the nature or extent of the work to be undertaken by the Engineer.

On appointment, it was thus to be expected that the Engineer would help to draft his own terms of reference. This is a skilled process which not only establishes his own duties but which also helps to define precisely what the Client expects to achieve as a result of the project, possibly in relation to other developments that would otherwise have been unknown to the Engineer. It has often been my experience that initial draft terms of reference received from a Client have described more how the Client would like the project to be undertaken than the essential feature of the principal objectives to be achieved. In the absence of such knowledge, better ways of achieving the desired objectives will be eliminated before initial planning has even started. Where there was a shared responsibility in preparing the brief, the Engineer would accept a considerable degree of moral responsibility for delivering the project in accordance with the needs; the Client the responsibility for the legal background to the undertaking, its finance, subsequent operation and maintenance. Where the Client was appropriately staffed, frequent communication and discussion helped to provide common viewpoints and continuity between these aspects. In the absence of this capability, the Engineer had to engage in a certain amount of diplomatic guidance in order to ensure smooth transition from construction to operational management. As a corollary, the first appointment with an initial task of determining what needs to be done, and how the appointee should contribute to its doing, cannot be based on price.

Among the several studies to be undertaken for the project, the site investigation was often the most vital. This required exploration of information held by departments of geological survey and acquisition of hydrological and similar data, where they existed, from available sources. In parallel, a strategy for construction would begin to evolve. In the play between the basic data about the natural background and the requirements of the project would emerge a strategy for its construction or possibly alternative strategies to be tested. This provided a key to the design of the site investigation. As the ground investigation proceeded, so could the strategy for the project become firmer. So, also, could expectations from further ground investigation be expressed more firmly, arousing alertness to any unexpected feature that might demand a change in interpretation. From the outset, therefore, the project evolved from interaction between several features, probably iterative, rendered impossible if these are separately commissioned.

When preparing Contract Documents for the project, the first essential is a clear definition of what is required, the quality standards to be met and any particular restrictions affecting how it is to be achieved. Absolutes should be avoided if these are not intended. Is 100% watertightness practicable or is it more to the point to specify what would be acceptable? Not only will this approach be helpful to a tenderer but it may well save legal argument if standards achieved fall below expectation. Furthermore, this will provide a valuable reminder of the need to determine the specific features behind the establishment of such a criterion. More generally, where uncertain features may significantly affect the choice of method or the cost of the work, there will be benefit in setting a limit to the range of conditions to be borne by, or specific assumptions that may be assumed by, the contractor, achieved through the provision of 'Reference Conditions'. These relate to features capable of measurement or of reasonable determination by the Engineer. The term 'Reference Conditions' dates from 1978 (CIRIA 1978) and has a rather wider field of application than the term Geotechnical Baseline Report (GBR). Reference Conditions may include stated limits of physical properties of rock and soil to establish limits of the Contractor's liability, as does GBR, but they may also address other features related specifically to the scheme of construction but beyond the Contractor's control, such as assumptions about geological horizons between boreholes. For the Øresund Link Project, the risk of icing of the sea was included, for example. Reference Conditions may refer to a range of values of a parameter, allowing a different unit cost to be priced for each subdivision of the range.

However, the prospective project might be advertised; it was then important to obtain information on which a satisfactory list of tenderers might be drawn, the means varying between Clients, respecting statutory duties or preferred practices. The Engineer knew that the elimination of an unreliable tenderer at this stage would be a great deal easier (and far more satisfactory for the Client, with a possible auditor breathing down his neck!) than the rejection of an unsatisfactory but lowest tender. Rules for accreditation of tenderers were necessarily drawn up prior to invitation to establish fair play.

A site meeting during the tender period fulfils several objectives: first, it ensures at least some familiarity with the conditions of the site and with particular features seen as relevant by the Engineer; second, it

Engineering, management and the law

provides an opportunity for the Engineer to appraise each tenderer; third, and equally important, it allows the tenderers to take a view of the Engineer as to competence and openness. The site meeting often provided the last chance to clarify an issue raised by a tenderer or, on occasion, by the Engineer himself as a result of the discussion. It is, of course, imperative that all tenderers be invited to any such meeting and that any individual query should be answered by the Engineer to all tenderers.

I have found a gulf between the views of lawyers and engineers in relation to the records of site investigation. To a lawyer acting for the Client, it seems that factual reports are made available principally in order to establish retrospectively that these could have been inspected by each tenderer, regardless of the grudging manner in which this might be undertaken. Although these may constitute the only reliable data of such nature, a disavowal of responsibility for their validity by the Client was not uncommon. To the Engineer, site data are regarded as a valuable resource for the project, providing a sound base for the design of the scheme of construction. It is in the general interest of the project that these are freely available during the period of tender. Interpretative reports, which may have been commissioned and are deemed relevant to the project, should also be provided, for guidance and without warranty. Those who prepared them will have special expertise and had longer to digest the site data than a tenderer; they will not know precisely how the tenderer intends to apply this information. The Engineer should not be put under the moral dilemma of having received warnings of a pertinent nature but unable to mention these to a tenderer (until the possible time for litigation). The object for good project administration must be towards a shared view on site data as the starting point of reference.

The completed Tender would be required to be accompanied by data concerning resources available for the project, including special plant and know-how, also to indicate understanding of the contract provisions, in order to allow choice between tenderers to be made on vital features beyond the monetary value of the Tender. The enquiry document would specify these features and indicate the general manner in which, by objective and subjective means, these issues would be taken into account. Where an innovation in design was introduced, or where there was any indication that a tenderer intended to adopt an unfamiliar approach, there would be need to ensure that the intentions were clearly explained and adequately agreed by all Parties before acceptance of a Tender.

Where Reference Conditions (or Geotechnical Baseline Report (GBR)) are included, there needs to be explicit provision for the additional cost. This may take the form of provisional items, whose quantity would be measured, to be priced by the tenderer. There may be a Provisional Sum from which the costs may be drawn, where measures that need to be taken require later definition. The other option is for a hidden contingency to be used to meet such costs. There is a cynical point of view that a contractor will find ways of swallowing any stated contingency and hence to conceal its magnitude, this occurs only with a weak Engineer. My own experience is that openness in all such features is beneficial in engaging trust and in obtaining, as a consequence, a fair bargain for the Client. In the 1980s, as another example of the misunderstanding of the potential uncertainties of construction, contingencies tended to be banned by Clients on advice from their financiers, lawyers or accountants. This had the effect of threatening the basis for cost adjustment and would introduce all the high costs associated in pretending that uncertainty did not exist. The Engineer tended to be criticised for expenditure of the hidden contingencies, although he had advised their inclusion as a means of avoiding additional uncertainty and cost.

The capability of the ICE Conditions to achieve a successful project may best be established by illustration of specific examples in a subsequent chapter. Success may be judged upon:

- Expectation of the Client more than satisfied.
- Adequate financial reward for Contractor, with encouragement towards best practice and innovation.
- Adequate financial reward to Engineer, satisfied by a project of good engineering, well co-ordinated across all its elements.

All these features need to be achieved without disregard for the interests of others.

The earliest traditions saw the project designer and supervisor, whether or not designated as 'the Engineer', viewed as the sole professional engaged in a commercial relationship between Employer and Contractor. This caricature belongs to a simple age. The project was defined, the Contractor provided labour, plant and materials to undertake the work in a traditional manner. As the task became more ambitious, so was ingenuity applied to the means for its achievement in plant, techniques and expedients. So, in consequence, did professional engineers become increasingly involved across all aspects of the execution of the project. In exercising his duties the Engineer needed, therefore, to be aware of the subtler needs to safeguard the legitimate interests of the Contractor. Programming of operations in construction may be interdependent to a degree that a relatively slight rephasing may impinge on other activities. The administration of a successful contract under the ICE Conditions requires the Engineer to have continuous regard for the interests of the Client, always compatible with an equitable attitude to the duties and expectations of the Contractor. If commercial attitudes prevail, this is an impossible balancing act. Without mutual confidence, success is unachievable. We are once again back to a question of trust, with the Engineer in a pivotal position to establish trust or to undermine its influence.

As a footnote, under the conditions of market economics, Adam Smith would encourage the view that mutual regard for honesty between the buyer and seller will help towards achieving good bargains. In construction, there is inadequate continuity between the Client and the Engineer, and sometimes between the Engineer and the Contractor, to establish such relationships, apart from exceptional cases. The cynics, and this category includes many lawyers, will have us believe that a successful project is an illusion, that the second best is to be aimed for, relying upon legal means to control the losses. It is one of my objectives to prove the contrary.

Risk procedures, discussed in Chapter 5, are now highly formalised. Previously, the Engineer would, to varying degrees of explicitness, have worked through a check list, drawing attention in the Contract Documents to those significant potential risks which remained unresolved at the time of Tender, indicating clearly who 'owned' the risk. It was then incumbent upon the Engineer, primarily in discussion with the Contractor, to foresee how a particular risk might first be manifested and the specific actions or reactions needed to contain it. The Engineer's supervisory staff had, as one of their duties, that of alertness to any feature that might be a portent of an unexpected event. The aim was that of anticipation in preference to retrospective attribution of blame and cost. The most testing time for the integrity of the Engineer occurred when, despite all such preparation, an event beyond the immediately obvious responsibility of the Contractor caused injury, cost or disruption. Was this wholly attributable to workmanship or did it have an external cause or one that might reflect on the Engineer's responsibilities? If the latter, the Engineer needed to ask himself why he had not foreseen the event. Why was the Contractor expected to have superior powers of foresight? It is worthy of note that, if the Contractor reasonably expects the Engineer to behave in a professional manner in such circumstances, the Client will obtain advantage through a Tender priced on this basis.

There is much current talk of 'escrow' in making available information on the structure of the Contractor's pricing policy for use by those concerned with the evaluation of claims, as if this were a new procedure. It is, in fact, familiar over many years to those of us who administered contracts through the ICE Conditions. Furthermore, this operation was undertaken by engineers with an appreciation of the extent to which costs could be affected by a change in circumstances, having regard to the degree of interaction between different operations and the scope for recovery from the disruption of an interposed event. Recent concern has been expressed about retention money being withheld by Clients (including principal Contractors for a project) who have become bankrupt; the setting of retention money into an escrow account might be a solution for this problem.

Part of the Engineer's function was to ensure that the Client was kept fully informed about all matters which might affect the programme or the cost of the work. Any problems needed to be discussed with the Client so that, if any action were expected from the Client, time would have allowed its preparation. In the event of cause for any form of settlement calling upon the Engineer's judgement, the wise Engineer would ensure that there was always a positive margin between the expectations of the Client and the Contractor. Planning of the hand-over between construction and operation needed to be undertaken thoroughly and early, particularly where there was to be a period of commissioning during which the training of operators would occur. As-constructed records and instructions from plant and control manufacturers would be provided as part of the works completion. The preparation of a project report for the future use of the Engineer's office is a valuable means for ensuring that experience from the project was added to the corporate know-how, to help in future estimating, in describing expedients and procedures and in recording the causes of problems. The preparation of this report provides excellent training for the young engineers concerned with the project.

Another positive feature of the work undertaken through ICE Procedures was that there were only three principal Parties involved in the construction project, the Employer (as Client), the Engineer and the Contractor. The development of mutual trust, upon which success so much depended, was in consequence more readily achieved than under the conditions of the present day with a growing proliferation of independent participants. In the most complex set of relationships, there may, for example, be a Proof Engineer, an Engineer acting for the long-term owner of a Private Finance Initiative (PFI) Contract, a Contract Design Manager, representatives of funding banks, Government and possibly others. The most remarkable newcomer is the external certifier for what purports to be self-certification by the Contractor. This proliferation makes for confusion, adds to cost and obstructs innovation. There remain more attractive options however.

The concept of Partnering is not new. The practices originated in process industries where long-term relationships developed between Owner and Contractor, exploited to mutual benefit. The benefits for construction, in the absence of the continuity of serial contracting, are less immediately obvious and need to be obtained within the individual contract – but perhaps even the word 'contract' needs revisiting. The project for Heathrow Airport's Terminal 5, for example, is based on Agreements between the Client, British Airports Authority (BAA), and the appointed organisations, who subscribe to the Delivery Team handbook. This is for a project depending for success on innovations in construction and operation, with strong interactions requiring interpenetrations of thinking. Partnering is more a state of mind than a specific formulation. Egan's notion of 'assembling the team' is essentially what is intended to occur, each newcomer expected to make professional contribution to the success of the project, as the primary objective. Rewards are intended to compensate fairly for the work, including unreserved share of know-how, optimisation of the whole dominating over separate optimisation of each part. Each participant will expect a form of bonus applied to project success overall, measured against established factors. A Target Contract may provide a basis for Partnering, with the proviso that uncertainties have been carefully evaluated, risk clearly allocated where control may be applied and the nature of circumstances leading to change to the Target Value set out fairly and unambiguously. In the past the ICE Conditions have frequently been applied with the essentials of Partnering, mutual trust. The Øresund Link and the Channel Tunnel Rail Link have each reaped the benefits of Partnering attitudes, if yet based on more traditional Conditions of Contract, applied in an enlightened manner.

4.6 The engineer and the law: pre-project
Statutory authorities are permitted to undertake certain classes of works without the need for express public sanction. Otherwise, projects may

be subject to planning consent leading, for a substantial proposal, to a planning inquiry. This may be a brief informal occasion to allow expression of views of those affected by the proposal. At the other extreme, the process may take years, being used in effect to decide major policy issues as well as the specific proposal. This is something of a parliamentary fudge, in that Government is relieved of the apparent responsibility for a policy decision unpalatable to some, although the ultimate decision remains that of the Department of Government to which the Inquiry Inspector reports. The case will be argued between lawyers who call upon expert witnesses to present written evidence on which cross-examination may be undertaken.

Alternatively, major projects may be promoted through a Bill taken through Parliament, a 'Hybrid Bill' if private interests are also affected. The engineer concerned with a project may find himself required to give evidence before the Committee (usually of the House of Lords) appointed to investigate the merits of the proposal. A report will be required for this purpose, supported by 'exhibits' in the form of diagrams and illustrations, on which the engineer may be questioned.

In all such encounters, the engineer, as expert, needs to make thorough preparation, with possibly a rehearsal before a 'devil's advocate', who will put the most awkward questions to the witness. The cardinal virtue is to present direct objective evidence supported by genuinely held professional opinions, not diverted in any way towards advocacy. Questions may well stray beyond the competence of the witness; he should not hesitate to admit this in declining an answer. His reliability as a witness will be fatally undermined by an unconvincing reply to a question on an unfamiliar topic, patently beyond the field of his expertise. Since the witness will be 'instructed' by lawyers, he will be subjected to more or less persuasion, during preparation for the event, to support a particular point of view. He needs always to understand that they are advocates, he is not. He cannot turn to them for support when he is under examination. As any case proceeds, it is difficult for the witness to maintain objectivity against the overriding climate of victory or defeat of all others in his 'camp' in their presentation of the case. This issue is further stressed below.

The expert witness has the primary function of presenting objective evidence to a tribunal based upon his professional knowledge and experience. The expert has secondary functions in advising his instructors on technical matters, and, under instruction, possibly representing their interests in respect of his expertise in negotiation or mediation. He should always be quick to react to new evidence that may cause him to revise his opinions. The opinions must be his own, except to the extent that these may depend on specifically referenced external sources or, for example, tests acknowledged as being undertaken by others.

4.7 The engineer and the law: the project and beyond

As a project proceeds, the possibility of dispute needs to be considered, the probability relating to the overall form of the contractual relationships. Disputes may be resolved internally or may require the services of external agents. The internal arrangements will depend on the nature of the contract, through, for example, the Engineer of the ICE Conditions or his less coherent counterpart of the NEC. Externally, there are many forms and procedures:

- an appointed Dispute Resolution Board (or similar title),
- a procedure for *ad hoc* Dispute Resolution by means described by the Contract,
- a form of mediation, whereby a mediator attempts to find an acceptable intermediate position between the starting points of the two sides carrying, as it were, a white flag to and fro between entrenched positions to arrange a cease-fire,
- direct procedure to arbitration or legal process.

I find great difficulty in reconciling the views in principle between lawyers and engineers as to the desirable characteristics of such intermediaries. Those who belong to both professions appear to veer towards the lawyer's point of view. If the project is a 'lawyer's contract', constructed on a strictly commercial basis without the intent of reliance upon professional judgement in relation to uncertainty, there is little prospect for the professional engineer's view prevailing in dispute resolution. For this reason, the activity needs to be looked upon as part of either the lawyer's or the engineer's style of contract described in Section 4.5 above. If, for example, the contract has been set up to satisfy the perceptions of lawyers of financiers or insurers, understanding only the nature of a 'commercial' contract, this preference will extend to all aspects of dispute resolution. One of the objectives of this book is to encourage an alternative view, in the interests, *inter alia*, of their clients – but the present concern is to consider the different scenarios.

The 'commercial' approach will expect dispute resolution to be undertaken principally by lawyers, with one or more engineers acting as assessors advising on technical aspects. This stage may be seen as essentially in preparation for the possible progression (or drift) towards arbitration, by a body such as the International Chamber of Commerce (ICC), where either Party may contest the findings. This is consistent with the overall legalistic approach to construction. The Channel Tunnel was one such example, where the experience of the Disputes Panel (Comité des Experts in French) described in Chapter 8, may be contrasted with that of the Dispute Review Boards for the Øresund Link between Denmark and Sweden. This was, unlike the Channel Tunnel, essentially an 'engineer's project'. All potential disputes were resolved without formal intervention of the Panels. 'They also serve who only stand and wait' (Milton, 'On his blindness'), as we are reminded. Of course, this contribution to a successful project needs to be kept in the context of a soundly devised system of an 'engineer's' contract.

There are, in fact, two totally different procedures. The legal approach starts from a defensively constructed contract to protect the interests of the Client. Each participant to the project has perforce to adopt countermeasures of protection, passing risk 'down the line' rather than managing it. Ensuing disputes are treated as forerunners to determination by a legal tribunal. This is an expensive route, one that condemns expenditure into defensive postures rather than innovation, and it is yet more regrettable in that the construction industry receives the blame for the consequences.

The Engineer's approach in dispute resolution, as elsewhere, is to address the capabilities of the project and help to prosper a system which operates with continuity and efficiency in addressing risks and opportunities. Problems along the way are addressed by engineers looking for solutions, at least cost to the project. There are numerous compromises between the lawyer's and the engineer's approach. The points of departure, however, are so different that it is scarcely surprising that many of the most serious and expensive problems have arisen from lawyers effectively taking over systems originally devised by engineers, without understanding the nature of the underlying premises.

The debate is not going to be resolved in an instant. There can be no doubt that, for the large projects which cannot be precisely defined at the time of tender, with high exposure to natural forces or where there are several interfaces between contracts and with influential external bodies, the engineer's approach is the only likely way to success. A major contract in Canada several years ago had been turned, by lawyers, into so complex a set of documents that only one man could claim to have read them all!

4.8 The expert witness

Where an accident occurs in a project, a prosecution may follow in the criminal court, for which the service of expert witnesses may be required. In comparable manner, a dispute, depending upon the structure of the contract, may proceed to arbitration or to trial by a Civil Court. So far as the Engineer as expert witness is concerned, procedures are generally similar and, in these days, correspondingly protracted. Different titles may be given to the reports or proofs of evidence that may be required. Having regard to the nature of the process, and the extent to which the result may hang upon the evidence, the prospective expert needs to carefully ensure that his credentials are understood by those who wish to appoint him and qualified by his own understanding of the issues (which may be insubstantial at this time). First, he needs to establish clearly his field of competence since it will be a cause for embarrassment if preparations go forward with the instructing lawyers expecting a broader coverage than can be offered. The cardinal sin of the expert is to trespass beyond his area of competence. If found out in this respect, the rest of his evidence will be tainted. He needs to be able to express his views cogently, coherently, concisely and always objectively. If his reports need editing by others, the risk arises of thoughts being introduced inadvertently which are not his own. He should use language comprehensible to his audience. This will vary according to the nature of the tribunal, be it a group of engineers, a group of engineers and lawyers familiar with the case, a court with a judge, or a court with a judge and jury (for certain classes of criminal prosecution).

Lord Woolf, Chief Justice, prepared a report in 1996 on recommendations for the improvement of procedures in the civil courts (Access to Justice Final Report 1996) which was followed by the publication of Civil Procedure Rules (CPR). These, in their turn, gave rise to a Code of Guidance for Experts prepared by a working party under Sir Louis Blom-Cooper QC.

A body in Britain which first met in 2000 – the Council for Registration of Forensic Practitioners – is principally concerned with medical evidence, but its remit includes, in principle, the wider group of all those who may serve to give expert evidence in legal cases, i.e. forensic practitioners by definition. Those who register will need to meet explicit criteria assessed against external standards.

Civil engineering in context

At the time of preparing this book, a particularly striking example of unreliable expert evidence has arisen as a result of the appeal by a mother, a solicitor, against her conviction for smothering two 'cot-death' babies several years apart. A pathologist, who changed his evidence on the cause of death of the second child from shaking to death to smothering, was supported by an eminent physician, who is also alleged to have made similar change in attribution of cause of death. More seriously, the pathologist had failed to reveal evidence of a respiratory infection. Additionally, the physician informed the court of the remarkably high odds against a second child in a family suffering a cot-death (this having been the cause of death notified for the first child, but for which no autopsy was undertaken). The advice was based upon the presumption that if the risk for one child was $1/n$, the risk for the second child was $1/n^2$, whereas, if every child has the same exposure to cot death, the odds remain $1/n$. In fact, little was known, at the time of conviction, of the factors that might contribute to cot death. It was possibly, later considered probably, wrong to eliminate genetic and similar factors which might affect the susceptibility to cot death, so the odds against a second child succumbing might be considerably less than $1/n$. A subsequent case of attempted prosecution of a mother of three children who each died in infancy, led to a general dismissal of the evidence of the same eminent physician when inherited susceptibility to cot death in this family appeared to be established. The physician appeared to have pronounced a widely accepted, but erroneous theory, that a second cot death in a family must imply a non-natural cause.

This example may seem far removed from engineering, but it illustrates the serious nature of a failure of a witness to present a fair view of his knowledge of the case, and of another witness failing to recognise his inadequate grasp of rudimentary statistics, while building a theory on insubstantial evidence. Where, it is pertinent to enquire, were the expert witnesses for the defence? There is a further moral from this experience concerning the notion of a single witness to the Court, discussed below. Comparable, if less personally tragic, instances are known to the author from the testimony of expert witnesses in arbitration, usually as a result of selecting evidence which supports a favoured theory.

It is necessary to emphasise the difference between the objectives of the tribunal and the function of the expert witness. The law depends upon evidence presented before the tribunal, with the burden of proof normally falling upon the claimant in a civil case, upon the prosecution in a criminal case (but not so where prosecution occurs under the Health and Safety at Work Act of 1974). The expert has the concern of studying relevant evidence, against the background of his own knowledge and experience, to arrive at the most probable explanation of the event in question. He usually finds several contributory factors, some of which may support one side, some the other. The expert needs to explain why the evidence that he presents supports his conclusions. It is not his function to declare one side or the other to be the 'guilty' party, since proof to a scientist or engineer has a different meaning from proof to a lawyer.

The engineer, entering the lists as an expert, should ensure adequate briefing on intended procedures and programme (almost certainly subject to extension). Legal procedures are expensive and long-winded. There are ways in which the expert can contribute towards economy and efficiency. There are equally ways in which his ignorance of procedures may add to the cost. On one issue, above all, he should establish the degree of confidentiality of his reports and communications. Are these privileged or not? He may otherwise inadvertently damage his Party's case in areas outside his particular brief. On first appointment, the expert may be asked to prepare an initial review, based on evidence selected and provided for the purpose. He should carefully identify in any report precisely what evidence has served in its preparation and, for his own notes, what other evidence he has seen – and, often later of great importance, where, in the innumerable files which will come his way, it is to be found. When the expert is subsequently instructed to prepare his evidence for the tribunal, in whatever form this may be required, he may often find the subject area has been confined to those aspects on which he has expressed views favourable to his 'side'. He should ensure that other issues on which his opinion is not required in evidence, but on which, on account of his experience, he may be expected to be cross-examined, are understood by those instructing him.

Where there is more than one expert, it is advisable to ensure that the relationships and terms of reference are understood. Is there a hierarchy or are there overlapping interfaces? One witness may be dealing with more general issues covered in greater detail by others. The expert must always make clear where he is expressing his own views and where he may be depending upon views specifically attributed to others. It is not his job to ensure consistency between experts, although he should draw attention to any significant discrepancy. Where differences persist, he should ensure that he has seen all the relevant evidence supporting those conclusions which differ from his own. In a recent case I found differences between my own evidence and that of another expert witness arose from the fact that, at an initial stage of a project, I would have undertaken a simpler approach than him to analysis in design. Our different backgrounds and experience created this, in the circumstances, relatively unimportant discrepancy but one that required explanation.

There is an increasing practice of 'without prejudice' meetings (i.e. no formal reference may be made to the meeting and its consequences) between experts with the object of reducing the area of difference on technical issues, and hence to save time and cost of the proceedings. The meeting provides a signal to the lawyers of issues that will not 'run' on the basis of the technical evidence. The obvious benefits that should result from such a meeting are frequently not achieved, for several reasons.

First, many lawyers do not favour a process, which may reduce their scope for arguing the case, and I have experienced several examples of interference. The most blatant followed a day's meeting between experts representing four Parties involved in an Arbitration, after a set of conclusions had been agreed and was ready for confirmation. It was then that an American engineer announced that his 'Attorney' had instructed him to agree nothing. It is as well to recall that, as a direct consequence, much further cost was incurred by the Party that appointed him, before the case was abandoned by them at the start of the hearing. This practice at the present day would probably lead to a formal complaint and retribution.

Second, following the meeting when the experts report back to the lawyers on the text of agreement, modifications are frequently proposed to water down the substance of the agreement. Unless experts have strayed unwittingly into the territory of law, these pressures should be withstood since they will, and do, result in failure of any agreement.

Third, and this is the most general failing in my experience, the meeting is poorly timed and structured. The timing may be too early to permit discussion in adequate depth or too late, if the experts have already undertaken work in preparation which might have been avoided or where it is only discovered at the meeting that the ground covered by each does not correspond.

Fourth, engineers as experts without adequate experience of their role, tend to be excessively defensive of their Client, forgetting that one way in which their Client may best be protected is by avoidance of cost and loss of credibility in supporting technically indefensible positions.

Engineering, management and the law

The simplest and most effective way to solve such problems would be for the experts to have a preliminary meeting to agree timing, agenda and representation for the definitive meeting or meetings, also to appoint a chairman from among themselves to control each meeting and to draft the conclusions. Good faith is an essential ingredient. The experts must understand that it is in their Client's interest to avoid technical discussion in areas that do not illuminate the case. The experts' meeting may provide a unique occasion to understand features that should influence their views that have failed to appear in their briefing.

All expert witnesses are, in effect, addressing the tribunal, although instructed by lawyers on behalf of a Party. It has been suggested that a way forward, to avoid the problems of conflict between experts where they might be expected to arrive at the same view from the same facts, would be to appoint one (or more) witness directly to the court. The witness to the Court might also chair meetings between experts and report on the findings to the Court as *rapporteur*. This procedure would more closely resemble the practice of European courts, depending on inquisitorial examination as opposed to the adversarial practices of the UK. Experiments in the US on Court Appointed Scientific Experts (CASE) have not been without problems, for example by lawyers insisting on seeing all the rough notes prepared by the CASE witnesses (Teich and Runkle 2000). There may well be a case for a technical assessor (as happens, for example, in some Inquiries) but there would be much floundering if counsel had then to present the technical case for each Party, without access to prompting from the expert witness. There are many opportunities for greater efficiency in preparation, such as a single set of files of evidence accepted as relevant, and better use of experts in winnowing the substance from the mass of irrelevance. Experts could often save considerable cost by being appointed early enough to advise on the substance of the allegations made on behalf of the plaintiff which determine the bounds of the dispute. In a recent arbitration, divided for its bulk into three parts, I advised the lawyers on my appointment that in my view the technical evidence for Parts 1 and 2 provided no support for their case. In due course, after much further cost, Parts 1 and 2 were thrown out by the Tribunal.

Where there is no conclusive reason for preferring one particular explanation of events, technical debate can help towards resolution of the uncertainties. Such technical debate does not occur in the tribunal where the issues are handled through Counsel with limited technical understanding and with a different set of objectives, but it should be a useful function of the meeting of experts. It deserves comment that scientific evidence differs from engineering evidence. As indicated by Lewis Wolpert (*The Independent* 16 July 2000) scientific evidence should be judged on four grounds:

- the testability,
- the error rate,
- the degree of acceptability by the relevant scientific community,
- whether the results had been peer-reviewed and published.

The one-off nature of construction projects and the scope for professional judgement limit the application of such principles but, where the engineer relies upon supporting evidence with more general application, there is a reasonable expectation for agreement between experts. Unfortunately, too many expert witnesses apply their experience acquired elsewhere to the specific issue, without adequate reflection on the differences, or without reliance on positive evidence to establish similarities.

As already emphasised, the expert witness has a variety of roles to perform and it is important that those who undertake this work in the future should understand what is expected of them in each role. Those who have experience as the Engineer (see Section 4.5) will be familiar with a comparable simultaneous undertaking of different functions. The practical experience of a potential expert, and the positions in which this has been acquired, should be thoroughly examined before he is appointed; it is here relevant to recall that the tribunal may have the power to accept or reject the appointment. It is also my view, based on examples to the contrary, that those in the academic world, without practical project experience at a high level, who serve as experts, should confine evidence to the field of their experience and not, for example, air views on what a Party to a Contract should or should not have foreseen or done at a particular point in time.

To those with an enquiring mind, to serve as expert is often to engage in unlocking the cause, or more usually the train of causes, of an incident. The primary cause is too often a result of an initial lack of understanding of the equitable allocation of risk, failure to equate powers with responsibilities or the failure to interrelate engineering disciplines. It is then frustrating to find that rules of confidentiality, particularly in arbitration and in cases settled out of court, prevent a wider dissemination of the findings. A cognate issue concerns the rapid investigation of accidents or near-misses so that the findings may be promulgated rapidly and effectively, clear of the leisurely timetable of the legal process; this is an area for study which I initiated, currently under exploration by the Royal Academy of Engineering.

5
Systems and design

5.1 Systems engineering

So much has been written on the subjects of design, project development and management, mostly by authorities in their respective arts, that it may appear presumptuous for a practitioner to add yet another voice. If all were well, yet another contributor would indeed be otiose – but all is not well. The essence of project design is that of synthesis through systems; it is this synthesis that is too often lacking, the synthesis between the creative thought and the effective action. Engineering is an activity which depends on a holistic approach; the boundary of the system, in the treatment which follows, includes the behaviour of people as well as the technical aspects. In order to attempt to find an adequately comprehensive, if compressed, set of criteria for success it is first necessary to explore the several strands of a project, then to discuss how these may be woven together to represent the pattern of the specific project, the subject of design.

In 1991 I was invited to take part in a seminar organised by the Ove Arup Foundation, *Education for the Built Environment* (Arup 1992). The issue was predominantly that of co-operative working between the several Professionals for the Built Environment (PBE). I was joint author with Dr Francis Duffy (later PRIBA) of the opening Paper, 'Society's needs'. We summarised the functions of PBE as:

- interpreters of requirements,
- guiders towards means,
- seers for future change and the co-ordinators of the elements for success,
- organisers of consequential projects.

We then reviewed the extent to which these functions were being achieved. Our tentative conclusion was that failures were attributable predominantly to fragmentation of the elements of project definition and the PBE, on account of allegiance to different professions and the intricacy of project relationships. We also attempted to find a formula for establishing an effective balance and co-ordination between the demands of the market and those related to the needs of society, defining our objective as developing the notion of an 'intelligent market'. The conclusions from any such analysis, and from the seminar as a whole, was for the greater interactive working between disciplines, with far-reaching consequences for professional education and training, for working practices and for the processes of procurement for projects. The notion of the intelligent market endeavoured to create enlightened and co-ordinated pursuit of the short-term and long-term interests of the customer in such manner to be beneficial to the private and public interest. The planning of the city of the future, with balanced high-quality provision for living, working, education, leisure, travel and so forth was seen as providing improved return for the market-led investment. We emphasised the symbiosis between the public and private interest and defined the intelligent market, without attempting to be prescriptive, as one which achieved this combination most effectively. Publicly funded investment would be justified as enabling investment for focused market-led projects which, in their turn, would contribute to local public revenue. We roundly dismissed the view expressed by the philosopher Roger Scruton in *The Times* (31 December 1985) of 'a country where property, profit and advertising dominate the collective consciousness, where law is fiercely adversarial and unscrupulous' as the only – and preferred – alternative to 'the centralised planning of the socialist state'. We recognised a more enlightened social contract between the public interest, the private interest and the professional.

Professionals working through the intelligent market, combining the interests of the public and private sectors, overcome the Manichean distinction between these sectors made by politicians. Thus, at the time of writing, the Chancellor of the Exchequer insists that only those who work in the public sector are capable of resisting the urge to sacrifice all to personal profit. Central Government of whatever hue appears to be infected by excessively intimate association with the City, on account of their importance as a major source of revenue (but not originators of wealth), where only those motives prompted by greed appear to be discernible from the outside. The intelligent market respects and encourages professional standards, combined with stronger local decision-taking and exposure to popular view,

Systems and design

operating hand in hand. This powerful combination would overcome centralist prejudices and improve, for all to see, the benefits from society from joint public and private sector operations. The intelligent market is wholly in tune with Egan's notion (Egan 1998) of the customer-led project, that is by the intelligent customer.

The notion of systems engineering is rapidly becoming accepted as the framework for success, diffusing from electronics across the full spectrum of the more analytic to the more pragmatic, with civil engineers as the last converts. Considerable impetus has been given to this broad acceptance by the concepts developed by my colleagues, in Halcrow and Bristol University, advanced in *Doing it differently* (Blockley and Godfrey 2000), building upon the processes of decision-making. It deserves comment that the title, taken from the Egan Report (Egan 1998), leaps forward in imagination from the concepts trapped in the confines of contractual relationships of the Latham Report.

Several years ago I found it necessary (Muir Wood 2000), since the notion of 'systems' appeared at that time to be unfamiliar to civil engineers, to broaden the scope of the familiar term 'design' to *design*, a term defined to include the systems engineering for a project. Times have moved on, and *design*, which had few supporters (some have suggested not many more than one), gives way, with some relief, to systems engineering. I believe that, nevertheless, it remains of value to set out my notion of *design* since this term continues to provide a common link (or cause of misunderstanding) between engineers and other professions.

In 1991 I was appointed member of a Steering Group on Architectural Education (RIBA 1992) chaired by Richard Burton, the other six members being architects, including Dr Francis Duffy soon to become President of RIBA. I approached the task with apprehension, foreseeing difficulties in aligning the views of an isolated engineer. I was soon disabused, discovering that, whatever our different points of departure, we shared a common destination. At the first meeting, I suggested that our professions were divided by different notions of design and that, since design was at the heart of our activities, a definition of design would help our debates. I assumed that the galaxy of architectural talent would leap to this challenge. I soon found that such a definition was expected from me; in due course, with considerable diffidence and apprehension, I suggested the following:

Design is the central element of the art of architecture – and engineering. Design should therefore form a central feature of education and training of the professions' proponents. But what is design? The word is used by different groups of people in different contexts. Thus, some architects equate design to depiction of form or style; some engineers equate design to calculation and analysis. The dictionary implies artistic and functional objectives but decouples these meanings and gives no clue as to the process of design. In the most general sense design denotes the continuous thread, the translation of ideas into achievement. In a restricted sense of design, abstract notions of form may be expressed by drawings and models. For professions engaged in the built environment, design extends the process through to ultimate expression in the form of construction (or planning) in town and countryside.

In order to discuss how design may be taught it is necessary to consider how it is achieved, that is to say the process of design. Essentially the designer starts with certain objectives which, at least in part, need to satisfy functional criteria. The designer has a reference store in his mind or an external data bank, giving access to more or less successful prior solutions to similar problems, with personal views based on such experience and on possible departures from previously perceived constraints. Frequently multiple criteria need to be satisfied simultaneously, with the need for stages of compromise in order to find the optimal solution. Achievement is only rarely measurable in a single unit such as cost so success of the ultimate product will be a question of subjective judgement. Only rarely will the initial idea be translatable directly to the end product. Usually the process will be one of trial and refinement, sometimes needing a restart on a radically different set of solutions. Often, the designer may need to consult the specifier of the criteria concerning utility to determine the practicality of modification or the prescription of more detailed nature to ensure conformity to the needs. There will also, except in the simplest examples of design, be the need for skills in different areas to be combined in order to achieve the product. As the design is evolving, therefore, each participant will to some extent be aiming at a moving target. Furthermore, the same overall problem seen from different angles of expertise may cause imaginative solutions not apparent to the initiator of the design process. While individual aspects of design of a complex artefact may be pursued individually to a certain degree, ultimately the scheme has to be considered as a whole. Only thus can the overall buildability and economic sequencing of construction operations be assessed.

The most essential element of design to be taught, therefore, is that it constitutes a system, conceptually represented as a series of iterative loops, the loops normally entailing communication between people or *between people and computers. Each participant will feed information, differentiating between fact and opinion in order to help to reach a conclusion with the minimum of reiteration. It needs to be emphasised that each participant will be performing parallel functions of reassessing ideas in his or her particular area of knowledge. It is necessary to regard each participant as providing not only constraints but, through ideas, opportunities for the others. For such a system to operate effectively there must be common 'awareness' between participants, to allow dialogue, with recognition of areas requiring the special skills of particular members of the team. Each must understand the expertise of others sufficiently to be able to address intelligent questions. Clarity of self-expression and self-criticism need to be encouraged, based on real projects which start very simple and, with experience, become more complex and nearer to those of real life. The student will be introduced to topics of aesthetics, art, science, technology, practice and management as appropriate to awakening of understanding, responding preferably to perceptions of need for such knowledge* [unpublished].

Somewhat to my surprise, this statement was accepted by my colleagues without amendment. I found this experience, which included attachment to a Visiting Group to a University School of Architecture, highly educative. Traditionally, the differences between the education of architects and engineers has been emphasised, the former often termed 'soft' the latter 'hard'. The architect starts by addressing the concept overall, with understanding of the constructional elements, the behaviour of buildings and associated applied sciences affecting safety, convenience and comfort added, more or less, as technologies that allow the concept to be achieved. His objectives, attributed to Vitruvius, are to achieve the combination of 'firmness, commodity and delight' or, in more familiar terminology: stability, utility and beauty. The engineer has traditionally studied individual sub-disciplines, i.e. structures, hydraulics and others, with little integration between the sub-disciplines and, until recently, little or no relation to broader issues. Many changes are occurring to these traditions, the architect becoming more familiar with the technologies and, through computer visualisation, able to manipulate scientific applications, of heat and light for example, without need to master the underlying analysis – which may of course introduce hidden dangers if operated beyond the range of validity. The engineering student of the present day, meanwhile, develops an ability for integration through studio-based design projects, also the wider areas of risk and optimisation which recognise no disciplinary boundaries.

Civil engineering in context

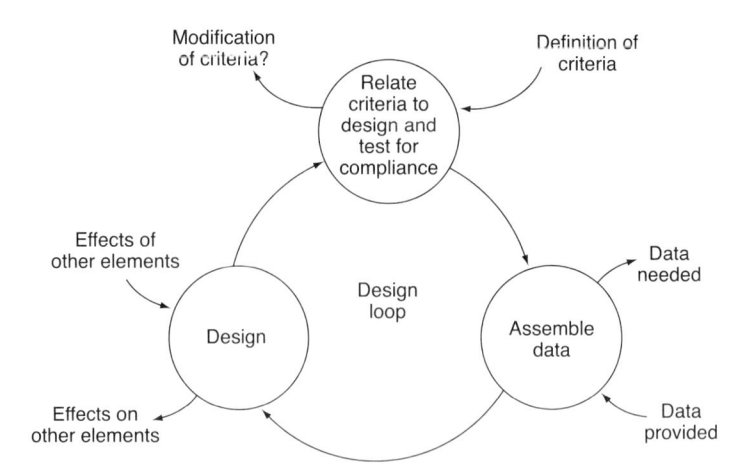

Fig. 5.1 Trial approach to design solution

In a simple linear system of design, a single set of criteria determines the design process and its outcome. More usually, even in the simplest circumstances, a set of trials of different hypotheses is needed so that the destination is reached through a series of loops (Fig. 5.1). For all but the most precisely defined objectives, the loops of trial and refinement of one element are related to similar procedures for other elements and for the simultaneous satisfaction of different criteria contributing to optimisation (Fig. 5.2). Each system of several processes contribute to the making of decisions, each aspect being addressed separately and then providing restrictions or newly perceived opportunities for other aspects tackled subsequently. Blockley and Godfrey

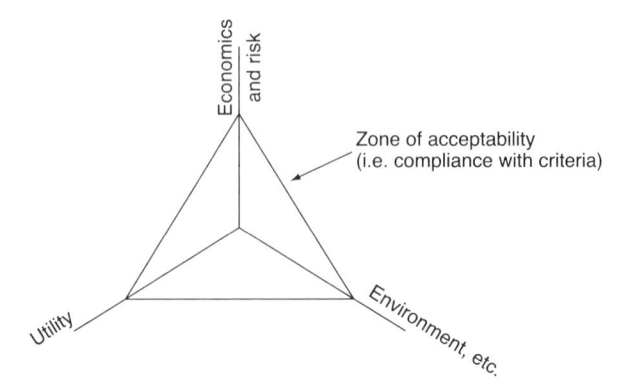

Fig. 5.2 Optimisation by simultaneous satisfaction of criteria

(2000) describe this system approach, providing the vocabulary of systems, sub-systems, cross-linkages, holons.... While the whole set of processes can be set out in a hierarchical fashion, there remain what are termed 'wicked' issues, where the rethinking in one set of sub-systems disturbs the conditions for another set of sub-systems, previously thought to be insulated from such development. An art of management is to determine orderly stages of decision-making, following the hierarchical principle, foreseeing the nature of interactions between sub-systems so that decisions taken at a lower level make minimum impact on those already determined at a higher level.

It is important to understand the demands on those who take leading parts in operating the system. In a simple linear approach to decision-making, the engineer has to find an optimal solution to a stated problem. This is a wholly technical matter satisfying simple criteria of feasibility, cost and time. His function is to apply his own professional discipline. The linear approach is economic in effort and is applicable where the terms of reference for a specific linkage of a wider system may be expressed with confidence and without the risk of diminishing thereby the alternative approaches which might call for recasting the terms of reference. In a linear system, the engineer accepts terms of reference dictated by others and, in his turn, contributes through his own conclusions to the terms of reference to be accepted by others who then determine more detailed consequences. There will always, even in such a linear approach, be internal systems at work in the mind as the engineer considers, rejects and accepts elements to contribute to his design. The engineer may well draw upon technical information and advice, possibly from colleagues, from references or from external sources, in the preparation of his conclusions, but these take the form of questions and answers, not as interactive processes. The engineer is working within his own discipline to which his task has been tailored and his communications are related expressly to the best application of his discipline.

In any complex system, where decision-making depends on satisfying a number of criteria, internally and externally controlled, success depends on the system addressing and taking charge of uncertainties. Decisions at every level are taken through the series of loops (Fig. 5.1) whereby information is interchanged between adjacent parts of the total system until the conclusion from each converges towards compatibility. The function of the engineer may then be simplified as operating in two alternating phases:

- The problem is formulated, reformulated and taken to a conclusion.
- The interactions with other aspects are discussed and measures agreed to test modifications towards convergence.

In this way, the engineer is involved in alternate periods of convergent and divergent thinking. These may also be represented as concentration on the solution of a specific problem with occasions of freer reflection, drawing upon past experience through the sub-conscious. This is probably an inherent characteristic of all those who engage in creative pursuits, expressed by the mathematician, Henri Poincaré (1845–1912) in these terms:

Discovery; between two periods of conscious thought, subconscious thought occurs. The subconscious or subliminal self plays a vital part in mathematical discovery [...] the subconscious role is as important as the conscious; it is not controlled automatically, it is capable of discernment, of selection, of refinement, it can choose, it has foresight.

... The selected phenomena of the subconscious, those capable of entering the conscious state, are those which, directly or indirectly, affect our perceptions most deeply. (Poincaré 1908 – translated from the original).

We can prepare a profile of the expectation of each one engaged in a system:

- Within his own discipline, 'expertise' is needed, the term defined by the level of knowledge appropriate to achieving an optimal solution.
- Within the areas of interest of those others with whom communication takes place and ideas are interchanged, in each process of the system, a second, lesser level of understanding is needed, adequate to question intelligently the views expressed in order to assess their cogency, relevance and adequacy of the options. This second level we may call 'competence', essentially extending 'expertise' to establish overlap between the different fields of expertise of others in the same and adjacent sub-systems.
- A third level of knowledge is required relating to aspects which may not figure in the processes of the participant but which may nevertheless form part of the total process of optimisation. This level of knowledge is referred to as 'awareness', helping to define the limits in the range of options within his own system.

As a simple example of these three levels (Fig. 5.3), the engineer exercising his expertise in designing a structure needs to display a degree of competence in discussions with the architect to establish

Systems and design

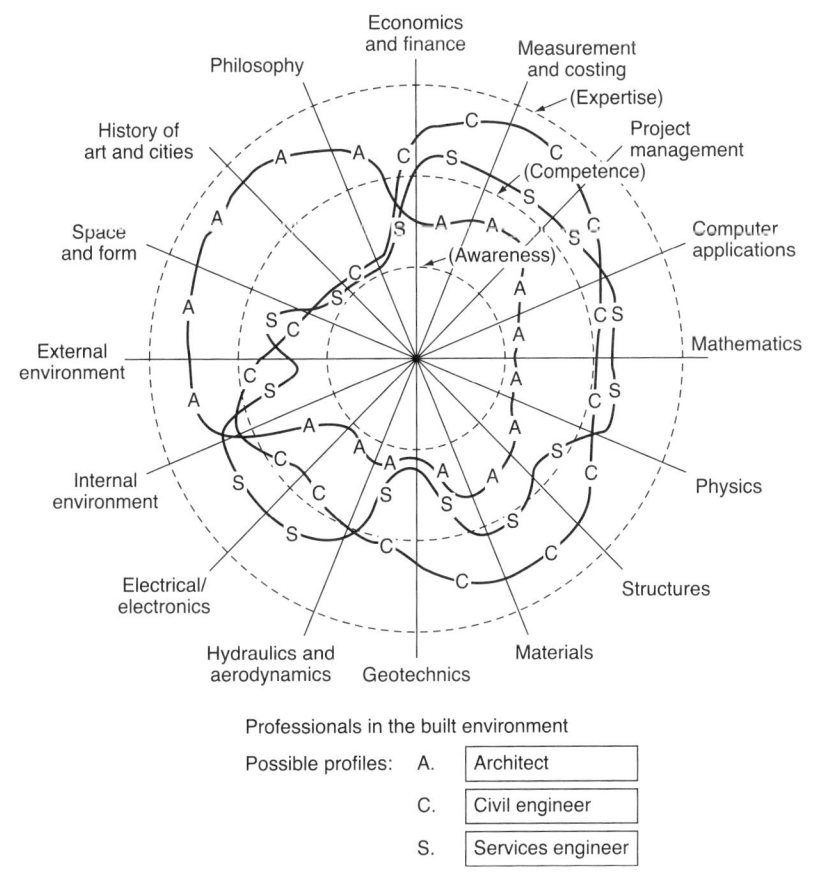

Fig. 5.3 Examples of levels of capability

the trade-off between the architect's vision and the structural practicability. He may also appreciate, i.e. indicate awareness of, external planning features which control the architect's options, taken as a 'given' in his own terms of reference. It needs emphasis that the definitions of awareness, competence and expertise must relate to the specific project; expertise for a simple project in a particular respect may correspond to competence in another more complex project.

The hierarchical structure in the operation of systems has been reduced to a pyramidal model of three levels in a number of quite different contexts. In a military context, these might be termed: strategy; tactics to deliver the strategy; and operations to deliver the strategy. In the context of the control of risk, this hierarchy is illustrated by a Cabinet Office Report (Strategy Unit 2002), reproduced as Fig. 5.4.

Civil engineering in context

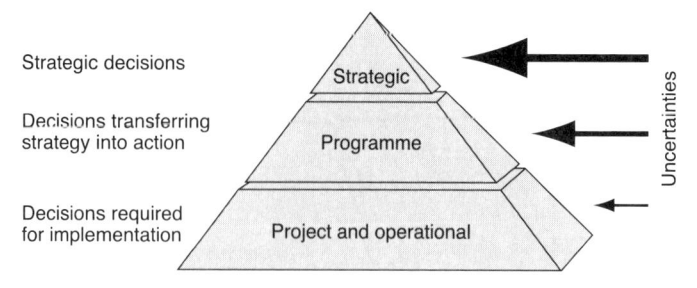

Fig. 5.4 Pyramid of risk control (After Strategy Unit 2002)

An illustration of this nature indicates the preponderance of uncertainty affecting the preliminary strategic decision-making processes, stressing the need that these be clearly set out so that the systematic development of the programme may set about their reduction and control. The 'project and operational' phase will have specific guidance on measures necessary to detect residual uncertainty and to anticipate and respond appropriately. What is missing in this model is any indication of interactions between the three levels to ensure the feasibility of tasks passed from one level to the level next below. The diagram reminds us of the importance of understanding the influence of external factors, in the instance of Government largely those of public understanding and acceptance, promoted by the concept of trust and by contingency plans for the unexpected. These features once again, in somewhat different contexts, occupy a universal place in any design project. It is significant that this report should have come from the Cabinet Office. There have been several recent examples of strategic Governmental decisions taken prior to consideration of the practicality of means necessary for their implementation at 'programme' level. Here we can instance the reaction to foot-and-mouth disease of cattle in 2001, several aspects of immigration control, the setting of numerous targets for Ministerial achievement and their subsequent abandonment, the unsolved problems of secondary and higher education.

5.2 Definition of the project

It should now be accepted that engineers of all disciplines recognise the need for a systems approach for efficient working in all respects, most especially in the identification and management of the components of risk. Systems imply continuity, purposeful continuity, in control and in decision-taking so that principles are not lost in the undergrowth

Systems and design

of detailed development. At all stages in the definition of the project we should be able to identify the conductor of the process, probably not the same individual throughout, but the succession so organised that the baton of continuity of thinking is passed on. The design proceeds coherently throughout the project, not in disjunctive steps, by duplication, nor through trench warfare between embattled groups defending unmarked frontiers.

From time to time a project encounters a hiatus. There may be delays in approval by a Client, a change in policy, problems in finance, or the not unknown, but always unexpected, prolongation of planning procedures. It is here that the maintenance of meticulous records justifies their cost. When the definition phase resumes, the steps reached in the process will be clear, including the degree of knowledge – which may well have changed during the delay – on which decisions or recommendation had been based. When data or criteria have changed, it is necessary to know when this occurred and to consider the extent to which prior decisions, of known date, need to be reviewed.

Estimates of expected project cost may be required from time to time. It is too easy to fall into over-simple models which imply that, by a certain stage in project definition, the estimate of cost should be capable of expression within a certain percentage margin of uncertainty. This doctrine takes no account of project-specific factors, which may be dominant. For a project wholly within a familiar tradition, relating to type, locality, technology and client, and allocation of risk, there may be no great problem of such an approach. Where circumstances depart from this degree of familiarity a more structured approach is needed. The engineer responsible will then need to define what is known, what is reasonably predictable and what remains at that time conjectural. Such judgements form a vital part of the project record, not only for purpose of review of the conclusions, but also for subsequent modification as any of the material factors changes. It is here that the lines of communication in the design process need to be kept particularly well connected. We can otherwise find a situation where conjecture, or possibly no more than speculation, is interpreted as prediction and, by a further step, leads towards certainty, because a weak link in the chain has failed to recognise the potential relevance of the feature.

As a simple example, a site investigation encountered what might portend an unexpected geological feature. This appeared to have no significance for the project at its current stage of definition; so the possibility was dismissed as unreliable data not meriting further investigation. Only too late did it become evident that this was not 'rogue' data but a real departure from expectation, introducing a new, unperceived risk into the project. The scheme of construction had meanwhile changed, greatly increasing the severity of consequences from the new interpretation of data.

Once we accept the principle of systems, the enlistment of co-operation between the operators of the system as an essential factor for success becomes apparent. Communication needs to be undistorted by concentration on special interests. The operation of the system takes priority over contractual procedure and precedent, which need in consequence to be subordinated to the system. A new age of partial enlightenment has already indicated the consequential benefit to all affected. An example of particular promise is to be found in the 'Early Contractor Involvement' scheme of procurement of the Highways Agency, which leads towards the principles of Partnering discussed in Section 4.5. What mainly yet needs to be achieved is a changed attitude of the bulk of those in charge of public and private decisions for development, brought up on standard economics or accountancy routines. Their dependence upon, and working towards, mutual trust, conforming to the intelligent market or the Treasury's notion of the 'enlightened purchaser', needs to replace that of apprehension and blame. These latter defects are immediate products of a 'fundamentalist' dependence on the free market, resulting from the concentration on self-interest of excessively commercial relationships. An essential feature of a system is that it only works effectively if all operators, of the processes that contribute to the system's chain, function consistently and coherently.

Manufacturing industry many years ago learned to consider the product and the process of its manufacture and marketing conjointly so that the system could be represented in simplified form as Fig. 5.5. This occurred at a time when the relationship, between individual engineering disciplines and the industry that each had traditionally served, was fast breaking down. In the construction industry, largely for historical reasons, the split between the process – the project design – and the product – the construction of the permanent work – has taken longer to heal. Increasing complexity of all aspects of the contributory processes drives the growing recognition of their interdependence. It is the purposeful continuity for each specific project which will fashion the appropriate relationships between participants, by contract relationships or other forms of co-operation. Only by understanding the degree of interactions between design and construction can success be achieved. The goal of success is thwarted in circumstances

Systems and design

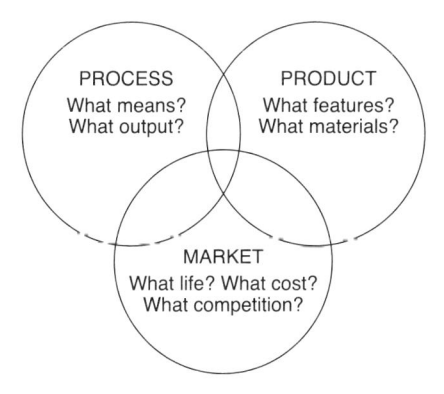

Fig. 5.5 Process and product for the market in manufacturing industry

where the Client, or his advisors, falls victim to established rubric which ordains that the structure of the Contract must come first. This is particularly reprehensible when co-operation between design and construction is deliberately thwarted to ensure full transference of risk to the construction phase. The project is then imprisoned in a straitjacket. One that starts in this manner with the law will probably finish there and all other interests will lose out.

5.3 Observation and anticipation

Observation has always played a prominent part in civil engineering. The early engineers had little else to rely upon. However protected the present-day engineer may feel by his confidence in conforming to interpretations of codes of practice and tradition, he should know that there is no guarantee that the particular combination of circumstances has been precisely foreseen by the codifiers. At least this was the general understanding and way of life prior to the intervention of QA. Subsequently, there has been an unhealthy expectation that, if the meticulous procedures of QA are followed, success is assured. This is, of course, nonsense. QA concerns a set of procedures in relation to perceived risk, and indeed sometimes introduces new and unnecessary problems in resolving issues of non-conformance. QA has nothing to offer on how to spot evidence for previously unperceived causes. The engineer remains just as exposed to the unexpected and just as reliant on his powers of perception and anticipation of their import and impact. Observation depends on training. Observation picks up the unexpected phenomenon, the curiosity, the early signs of a potential problem. At a more organised level, these may be departures from the expected trend of readings from some form of instrumentation. Anticipation looks ahead to the possible need for action to avoid unacceptable development of the condition. Monitoring implies a pre-arranged set of responses to observations, of a formal nature; in these circumstances the observation and the anticipation are coupled together. This combination leads to the Observational Method discussed below.

In general terms, familiarity with the system in which the engineer is working provides a back-cloth to the recognition of the unexpected. The system then requires a new 'loop' in order to restore an acceptable outcome. Often the observant engineer is not looking for anything in particular, simply exercising a sensitivity to exception from the expected, at all times receptive to the possibility that something may not be as it should be. Obviously, the earliest that such suspicion occurs in an unexpected set of circumstances, the better. The simplest way to demonstrate this state is by example.

London Clay spoil from the Potters Bar railway tunnels (Terris and Morgan 1961) and cuttings was conveyed by tipping wagons hauled on narrow gauge (Jubilee) track – probably the last time that this method was used for a major project in Britain. The spoil was tipped, spread and lightly compacted by bulldozer in a field designated for this purpose. The field sloped fairly gently towards the railway. During a periodical visit to the project, cracks in the surface of the spoil tip, transverse to the slope, suggested that down-slope creep might be occurring. Survey pegs soon confirmed this movement and established the rate, and the rate of acceleration. Cores established the water-and-air-filled voids ratio and hence the likely equilibrium liquidity index:

$$I_L = (w - P_w)/(L_w - P_w) \tag{5.1}$$

where w is water content, L_w and P_w liquid and plastic limits respectively. Reference to a paper by Skempton (1959) enabled the approximate undrained strength of the remoulded clay to be estimated, corresponding, for the geometry of the tip, to a factor of safety of little less than one. There was thus a potential problem, threatening not just the spoil site but also the railway. The minimum size of a hardcore 'dam' to stabilise the tip, and its tactical position ahead of the creeping toe, were determined. Meanwhile, the finished surface contouring of the tip-site was varied appropriately and agreed with the planning authority (fortunately with a fast-moving authority). The movement was stabilised as designed. The lasting consequence is a somewhat humped contour which could exercise future archaeologists.

A second example comes from the Clyde Tunnel (Morgan et al. 1965). This pair of road tunnels was partly in rock of the Upper Carboniferous, partly in glacial and more recent deposits. Generally, glacial drift as boulder clay immediately overlay the Carboniferous sandstones. Sheet piling was driven to rock-head to form a coffer-dam for construction of the north portal of the tunnels, which also provided the foundation for ventilation plant room and tunnel control building. During a site visit, following a few weeks absence, when excavation was proceeding under water (to reduce net ground load on piling) and approaching formation level, air bubbles were seen rising through the water. Their source was immediately recognised as the face of the approaching tunnel, under construction in compressed air. A layer of sandy silt was known to overlie the boulder clay. This pointed to absence of integrity of the clay across the area of the coffer-dam, relied upon to provide a seal to the base of the excavation. Excavation was stopped and the coffer-dam kept flooded. Pre-Contract boreholes on diagonal corners had indicated several metres of clay; boreholes promptly sunk on the other diagonal disclosed evidence of erosion of the clay by an unsuspected post-glacial stream. To achieve stability, the water-table around the coffer-dam was depressed by pumping from the underlying sandstone. An attempt to intercept the compressed air travelling through the ground, which might interfere with the drainage, was too successful. A borehole sited to strike a coarser, higher hump in the sand layer caused a powerful escape of air, bearing sand and pebbles, and was closed down. Special care was taken in excavating where the clay had been eroded. Once the potential problem had been explained, the appropriate counter-measures were evident.

A third example describes an occasion where observation occurred too late. This concerns a hydro-electric project in Fiji whose inclined shaft, linking the low-pressure to high-pressure tunnels, had encountered unexpectedly unstable rock. Several conjectural explanations had been advanced, including that of high horizontal pressures by a member of the consultants' board. A colleague, an engineering geologist, asked to provide an expert opinion, discovered that no qualified geologist or engineer had actually examined the shaft, on account of the extent of fumes caused by the Alimak diesel-driven climber used for shaft raising. The unstable rock was causing continual damage to the ventilation trunking. My colleague armed with video camera mounted the unfinished shaft atop the climbing shaft-raiser. He revealed that the predicted geological structure was disoriented; in consequence the shaft had followed a shattered basalt dyke. The conjectures, lacking observation, had meanwhile led to arbitration.

Observation can take many forms. A section of beach on the coast of the Dominican Republic was eroding unexpectedly. From a steep angle, the clear water allowed the seabed to be discerned to a considerable depth. Inspection from a light aircraft revealed damage to underwater coral reefs, allowing escape of a sand stream, whose course could be followed to deep water.

On occasion, a problem posed to the engineer may be solved by nature, where experiment and observation could entail irreversible effects. A linear lake parallel to the Honduran coast had previously been open to the sea about mid-way along its length. Would reopening of this silted channel, desirable for environmental and economic reasons, cause damage to villages on the spit near the old channel? This seemed improbable but the risk could not be totally dismissed. While reflecting, in some indecision, a violent storm broke through the spit and reopened the channel. Engineering risk had thus been transferred to a natural process – and no considerable damage occurred.

Another experience of maritime observation occurred for a beach in Sithonia in north-west Greece (Thessalonica). Here (Muir Wood 1970), the practicality of a marina project depended on the stability of a beach around a bay and the absence of appreciable long-shore drift. Approach by the Client to a professor of hydraulics had previously drawn a reply to the effect that a hydraulic model would be needed with at least one year's data of waves and currents; the delay was unacceptable. However, an initial study of a hydrographic chart of the bay suggested a stable concave shore between headlands, fed by a seasonal river towards the middle of the bay, its mouth not notably deflected. A site visit confirmed this impression, noting in particular the distribution of fine gravel, grading from 5–10 mm size, corresponding to that of the river bed. The hypothesis that the beach was stable (i.e. not dependent on longshore drift) derived from material, fed intermittently from the river, required certain conditions for the size distribution of gravel and sand, which formed the beach, to be satisfied. The material should be coarsely graded on the upper beach, the mean size diminishing from the centre of the bay towards the headlands; the lower beach would be expected to comprise fine sand only. These criteria were proven by samples obtained by a diver and, incidentally, soon thereafter confirmed by the discovery of a fifth century BC pot from a short distance inland from the beach.

5.4 The Observational Method

When observation is introduced as a specific process in the design system, the system may be termed the 'Observational Method' (OM) or Observational Design, the latter description emphasising that it is an element of design. Observation has been adopted informally as part of the design process by engineers over the centuries; modifications made during construction bear witness to this fact. For example, the spreading walls of the church of Santa Maria del Ser in Santiago de Compostella were evidently stabilised during construction by massive buttresses, as immediate responses to unforeseen foundation problems.

The OM has been systematised by Peck but in an unnecessarily, and possibly inappropriately, detailed manner (Peck 1969). The essence of the OM occurs where a safe and economic solution may be found to a problem such that an initial solution, to a selected standard, is monitored to indicate whether it may require to be augmented in order to be adequate. There needs to be a premeditated scheme for providing such augmentation during the time available, that is before any irreversible damaging action has occurred. Muir Wood (2000) has illustrated how the level for the initial solution depends on the relative cost of work undertaken initially and that which observation may indicate to be undertaken subsequently. The higher the relative cost of the latter, clearly the higher the level set for the initial solution. OM most readily lends itself to temporary works of ground support, e.g. the primary support in tunnelling or the bracing for excavations. Powderham (1994) has described a useful flexibility in the use of OM by the notion of progressive modification, utilising information gained during construction to work towards the optimum level of design within acceptable limits of risk, the final step corresponding to Peck's response to observation. This approach is particularly applicable to temporary works, where the degree of ground support may be provided incrementally. An example of this variation of the method has previously been described by Kidd for the primary support of the Orange-Fish Tunnel (Kidd 1976). Figure 5.6 demonstrates how early measurements of tunnel convergence were used to predict whether support by sprayed concrete or by sprayed concrete and rock-bolts was likely to be required to establish stability prior to lining. Examples

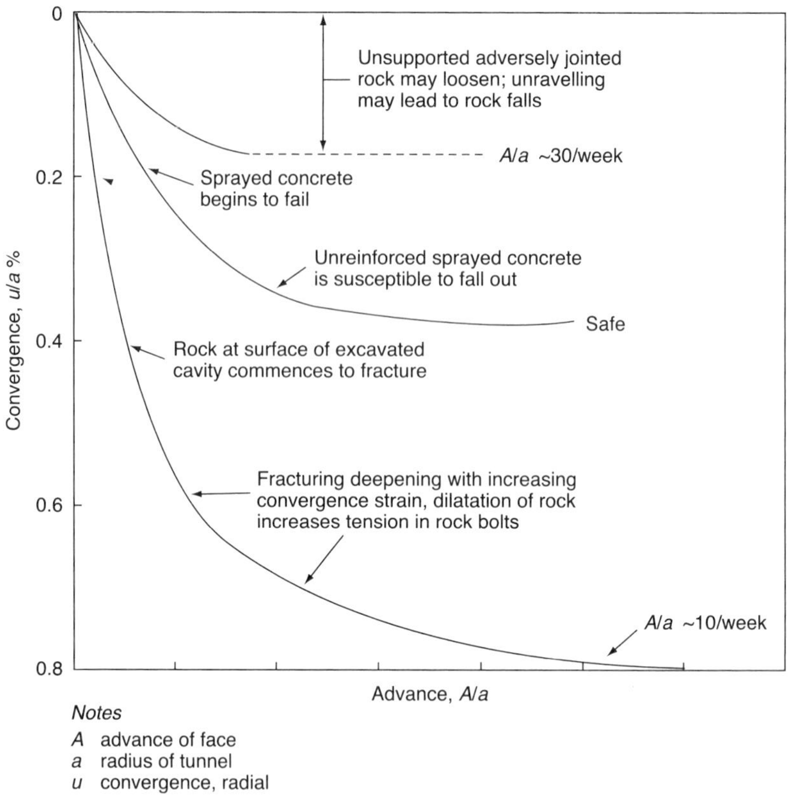

0
0.2
0.4
0.6
0.8
Convergence, u/a %
Unsupported adversely jointed rock may loosen; unravelling may lead to rock falls
A/a ~30/week
Sprayed concrete begins to fail
Unreinforced sprayed concrete is susceptible to fall out
Safe
Rock at surface of excavated cavity commences to fracture
Fracturing deepening with increasing convergence strain, dilatation of rock increases tension in rock bolts
A/a ~10/week
Advance, A/a
Notes
A advance of face
a radius of tunnel
u convergence, radial

Fig. 5.6 Tunnel convergence guidelines (After Kidd 1976)

of such application emphasise the need for purposeful continuity across the design and construction processes.

The criteria of adequacy for the initial solution (and for any intermediate solution) need to be clearly stated, accompanied by details of monitoring by instrumentation or otherwise. These conditions are often referred to as 'triggers'. There may be an 'amber trigger' to give warning of the need for more intensive monitoring; a 'red trigger' will indicate the need for counter-measures of a predetermined form. Planning needs to ensure that at any moment, the recording of a 'red trigger' and the appropriate response are accomplished within the critical preordained period, taking account of all the procedural processes intermediate between the observation and the execution of the supplementary work. Caution will be needed at the start to understand what this critical period should be. The question of triggers may be illustrated by reference to the monitoring of displacements of a primary ring of

Systems and design

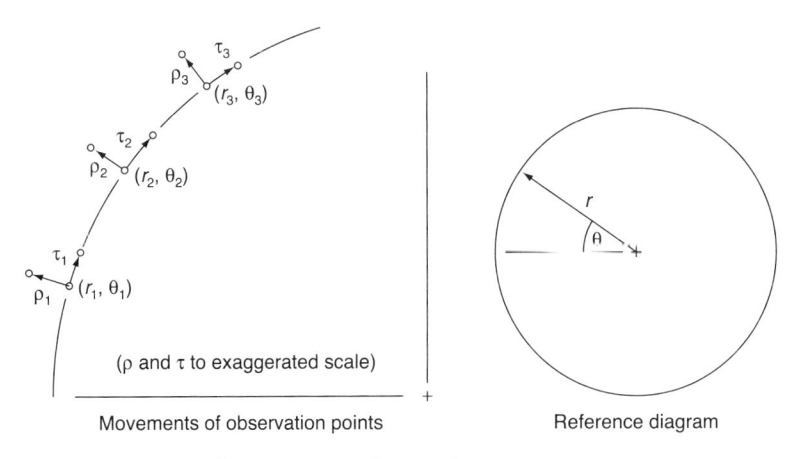

Fig. 5.7 Monitoring of primary ring of sprayed concrete

shotcrete as support for a tunnel (Fig. 5.7). From the viewpoint of integrity of the ground, inward movement (convergence) may provide one set of criteria for the red and amber triggers. From the viewpoint of integrity of the shotcrete, the criteria are somewhat different. Consider three adjacent survey points (1, 2, 3...) around a circular ring, at points (r, θ). If radial and tangential deflections are indicated by ρ and τ respectively, compressive strain around the ring between points 1 and 2 will be represented approximately by:

$$[(\rho_2 + \rho_1)(\theta_2 - \theta_1)/2 + (\tau_2 - \tau_1)]/r(\theta_2 - \theta_1) \tag{5.2}$$

and likewise for other sectors, while bending strain caused by change in curvature at 2 may be represented approximately by:

$$4[2\rho_2 - (\rho_1 + \rho_3)]/r^2(\theta_3 - \theta_1)^2 \tag{5.3}$$

and the monitoring system may be programmed to record these values. If such simple criteria for recording convergence had been applied to the Heathrow Express tunnel collapse (Muir Wood 2000), incipient failure of the tunnel invert, indicated by apparent compressive strain, at least an order of magnitude in excess of acceptable values, would have been evident several weeks before collapse occurred, allowing ample time for remedial work.

Coastal protection at Barton-on-Sea depended on the development of natural bays, with beach material retained between artificial bastions extending well beyond low water (Fig. 5.8). These bastions were formed by rubble mounds of rock-fill stones, with the seawards heads retained by spaced timber piles. At the toe, around the perimeter, precast

Civil engineering in context

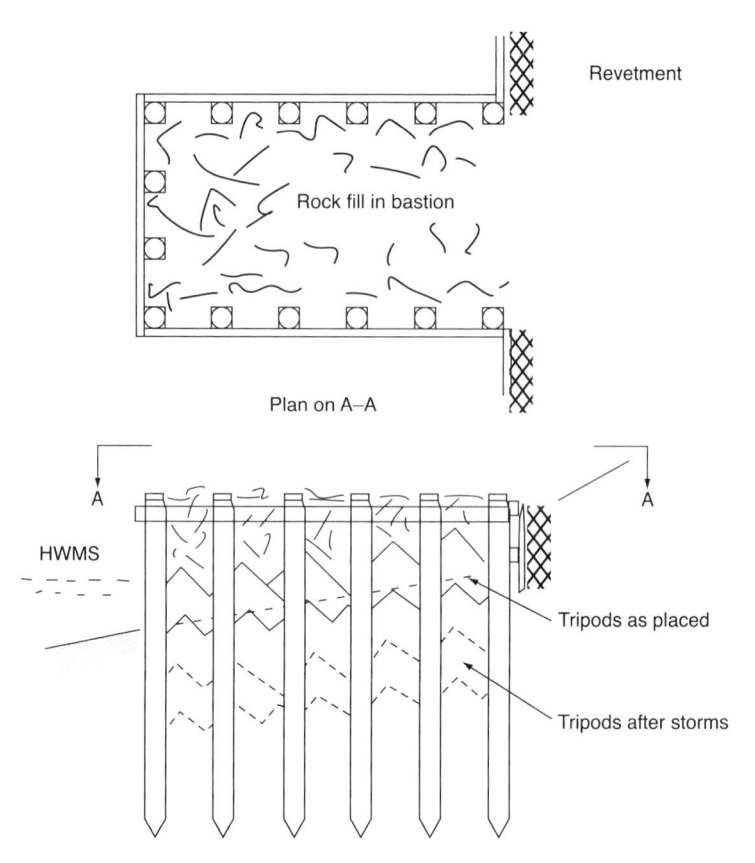

Fig. 5.8 Sketch of bastion at Barton-on-Sea

concrete Tripod units (2 such units form a cube, for easy 'breeding') were placed against the piles. These, with the remainder of the rock, were designed to settle vertically under the action of storm waves during the first few years. Periodical observation indicated when replenishment was required to maintain the feature, in effect using sea action to create a sound foundation. This represents an extreme form of the OM, since it was certain that the initial provision would be inadequate. The supplementary work of burying the foundation, on the other hand, was undertaken by nature

It has also been claimed that the OM may be applied by making excessive initial provision, which may then be reduced as a result of observation. There is here a logical confusion. Imagine a long trenched excavation in a uniform soil. The struts between sheeted side-walls for an initial length are over-sized and instrumented. As a consequence, reduced strutting is used for the subsequent lengths of the trench. Undoubtedly, this is an observational approach, but the OM, by definition, will only be involved if, at all times, monitoring provides for the reduced strutting to be augmented if need be. The initial length in this case serves as calibration for the OM proper.

An essential feature of successful application of the OM lies in a unified control of the operations concerned, which may readily be achieved through intelligent application of the ICE Conditions, the NEC or by a form of Partnering. It will be impossible to enjoy the benefits of OM – and these may be appreciable – where no system connects design with construction. OM may also be thwarted by external Proof Engineers, or those acting in comparable capacity, charged to oversee the interests of other parties. These external assessors may tend to press for the adequacy of each activity in construction, so that they can 'sign it off', whereas OM will recognise the possible need for supplementary work after the initial action. Where excessive responsibility has been thrown on the Contractor undertaking the work, there will be no apparent financial interest of other parties in encouraging the adoption of OM. There must always be benefit in undertaking a high standard of monitoring to expose the unexpected, whether or not the OM is adopted. The OM provides a desirable incentive for this activity.

The OM is given a bad name by those who claim to be working in this fashion but without exercising adequate rigour. A recent major tunnel project depended on achieving a high standard of watertightness to avoid damage to property by lowered groundwater table. The decision on the means to achieve such standards was delayed while tunnelling progressed, in the absence of adequate assessment of the stringency of necessary controls, on the basis of claiming that this constituted a valid observational approach. Problems arose on account of the absence of reliable estimate for the worst case, as a result of which no adequate contingency plan had been evolved to satisfy the principles of OM. High costs were incurred in retro-action and in the costs of permanent recharge.

5.5 The use of models

Representational scale models have been used throughout history by architects and engineers to demonstrate the nature, the appearance and, occasionally, the behaviour of their proposals. All engineering design makes use of models, in that the artefact is replaced by a model of physical properties, usually simplified, seen as relevant for the analysis,

Civil engineering in context

while other aspects are ignored. Another example, widely used at the present day, concerns the replacement of the ground by a ground model, with zoned simplified physical properties within regions defined in three dimensions (Muir Wood 2000).

The use of models to replicate behaviour of the prototype requires understanding of the physical laws in relation to scale. Essentially, a physical model is being used to close a syllogism of the nature:

$$a : b :: c : d \tag{5.4}$$

where: a represents analysis related to the model
b represents observed behaviour of the model
c represents analysis, comparable to a, applied to the prototype
d represents predicted behaviour of the prototype under the modelled conditions.

There are, for example, familiar accounts of Robert Stephenson's model experiments for the tubular bridges over the River Conway and the Britannia Bridge across the Menai Straits. Here, in fact, the 400 ft (120 m) Conway Bridge served additionally for intermediate observations for the Britannia Bridge, 1511 ft (460 m) continuous over four spans. Sir Marc Isambard Brunel undertook experiments on models of reinforced brickwork within the Thames Tunnel in the 1830s during a protracted interlude in the course of construction. If all material physical factors scale linearly, models may provide valid and directly applicable data, always taking account of the differences in construction processes (e.g. rates of cooling and drying out) and control of workmanship. For the direct comparison of geotechnical models, therefore, it is important that ϕ', the effective angle of internal friction, is constant for the range of loading and that c, the cohesion intercept, is zero. Since gravity determines the stress levels in normally consolidated soils, the syllogism of Eqn 5.4 may be bridged by the use of a centrifuge which generates radial acceleration $r\omega^2$, say αg, where $\alpha = r\omega^2/g$. If r, radius, is, say, 5 m a value of ω, angular velocity, of $\sim$13.5 radians/s (about 130 rpm) will realise a radial acceleration, αg, equivalent to 100g. Time for consolidation will scale as αd^2, where d represents dimensional scale of the model.

For physical hydraulic models, with a free surface, problems immediately arise since wave velocity scales in relation to $h^{\frac{1}{2}}$ where h is depth of water, explaining the need for vertical distortion and the expedients then needed to increase roughness of the wetted surface of the model if friction and turbulence are to be simultaneously modelled through Reynolds and Froude numbers.

Systems and design

In principle – and at increasing cost with degree of verisimilitude – numerical models may represent the prototype to a limitless degree by reducing the spacing of the grid or the size of the elements whose physical characteristics are used in formulating the constitutive equations of the material modelled. The numerical model thus directly represents the full scale of the prototype. The model code represents the virtual (as against real) representation of d of Eqn 5.4. The basic questions of every numerical modeller's self-catechism are, or should be:

- What do we want to know and to what degree of accuracy?
- To what degree of certainty do we know the characteristics of the prototype to be modelled?
- Does the model correctly reproduce the order and history of the several phenomena to be modelled?

Dealing with these features consecutively:

(i) There is a strange fascination, a form of macho instinct, to engage models of unnecessary sophistication which do not yet reveal the specific features of greatest concern to the engineer. If one or more characteristic cannot be ascertained within a range $\pm\delta\bar{A}$ where $\bar{A}$ is a mean value, a simple initial analysis will indicate the limitations on the reliability of the print-out from the model. The greater the complexity, the greater the cost – to a high exponent. To what purpose? The files of printout, to several significant figures, and the vividly coloured illustrations, carry an insouciant air of verisimilitude, beguiling to the innocent. There is at work a false objective to impress, rather than inform, by using a more complicated model than that of a competitor, all of no benefit to the Client, or to anyone else.

(ii) The verification of a model assures us that, within the stated conditions, it will provide a correct solution for the assumptions made. This may be achieved by modelling a problem of known analytical solution. If the model is being used for conditions beyond its intended scope, or if special features of finite elements, of tailored properties for example, are introduced which may affect the results, then verification for these conditions should be repeated. For a practical example, the shrinkage gaps between mass concrete and rock were represented by special elements; unwittingly, the modeller had the effect of turning the mass concrete into a prestressed slab whose behaviour was fundamentally different, but the change was unremarked by the modeller. The result was a

Civil engineering in context

serious failure, readily foreseen by a simple analytical approach to the problem – literally back-of-the-envelope.

(iii) Validation is the process of ensuring that the processes, and their interactions, to be reproduced by the model are correctly understood and modelled. For a model of non-linear stress/strain relationships involving a hysteresis loop between stress and strain in any loading and unloading cycle, each increment of change is irreversible. The solution is therefore unique to the particular order of loading and unloading of each element of the model. How well do the assumptions of the model in such respects correspond to construction? Would the use of a simplified elastic model, tested for sensitivity to a range of values, be adequate for the purpose and avoid introduction of errors and uncertainties difficult to diagnose?

(iv) The modeller must be aware of the critical features of the problem to be modelled. What is its purpose? Is there a simpler analytical approach which can more readily be tested for sensitivity to variations of uncertain factors? Does such a model, even if too crude to provide an acceptable solution, indicate the most critical combinations of the uncertain factors for each of the different aspects to be investigated?

(v) There is some confusion concerning the engineer's need for the command of mathematics, in an age in which much of the complexity of calculation is undertaken effortlessly by computer. This question is discussed further in Chapter 7. Two separate aspects need to be distinguished in relation to numerical modelling. On the one hand, concepts of physics will continue to be represented by equations, mostly differential equations, and the engineer will need to continue to understand how those which contribute to his technologies are derived; on the other hand, solutions of the equations by computer will continue to need to be accompanied by the ability to formulate the equations for analytical solution of simplified examples, to test validity and sensitivity to variable factors.

The chartered engineer should understand the nature of the problem solved numerically by the model and the manner in which this is undertaken. Modelling can generally be accepted as a mature technology. Detailed understanding of the code for a well-used model is needed, therefore, only by the engineer writing new programs, the numerical analyst, and those who identify and cure program faults.

Systems and design

Engineers should always recall that numerical models solve equations based upon scientific theory of a precise nature. They provide no information on the degree to which the scientific abstraction, based upon a set of axioms, represents the engineering artefact to be modelled. In products subjected to rigorous quality control, the degree of correspondence between assumptions and reality may be assessed with confidence and included in any application of the model. Workmanship, on and off site, and the properties of natural materials are site-specific factors whose influence on critical assumptions must be understood and allowances made. Structural models often assume simultaneous and instantaneous creation. Geotechnical models usually depend on time-dependent processes but rarely are these related in detail to the timing of the processes of construction.

When tackling an unfamiliar problem or where there is no simple means for optimisation through the use of a numerical model, the engineer should always consider a simple analysis, scoping a possible solution, i.e. providing the range for trial, and a basis for sensitivity testing, by identifying the principal factors. This process should be considered an essential part of any calculation. Analysis should be kept simple or the complexity obscures the objective. The engineer may use a simple force diagram, simple differential equations such as the examples elsewhere in this book, the use of area moments to check a structure, including the use of the identity:

$$\oint (M/EI)\,ds = 0 \qquad (5.5)$$

for a continuous framework. Each sub-discipline of engineering has its own means for undertaking simple first estimates.

In tunnelling, the conditions around the face of a tunnel call for a three-dimensional approach. From observation, for a particular set of conditions, this has by some been achieved by reducing the modulus of the core to be excavated to a hypothetical value to fit the data. This is a plain fudge; it does not consider the contributory factors and is hence unable to be adapted for any change in conditions, such as different support or different rates of advance. Simple analysis provides a better model, one which can be so adjusted.

The prediction of sea sediment movement by waves and currents provides an early example of interactive numerical models. The movement of beach sediments, predominantly by sea action, was roughly predictable in the 1960s by simple analysis, which could be improved by numerical modelling. I was also aware at this time that sediment movements offshore would require interactive models, one model relating waves and sea currents to the bottom topography, the other relating the rates of sediment erosion, transport and deposition to the same topography. By using the latest output of one model for the input of the other, the consequences of a selected set of initial boundary conditions, and possibly changing conditions of wind, wave and tide, could be modelled over time. No such interactive iterative models existed at that time. The practical application of such models occurred in the 1970s in connection with the design of cooling water works for a coastal power station, to be constructed first for an output of 1000 MW, subsequently to be extended to 4000 MW. A colleague, Dr C. A. Fleming, developed the first programs in attachment to the Mathematics Department of Reading University, while the initial planning for the station proceeded. Calibration of the first set of models was achieved in relation to the effect of storms as the protective breakwaters for the intake works were being extended. Subsequently, the arrangement for the first stage provided a large-scale physical model for the finer tuning to the predictions and design improvements for the second stage. These numerical models have continued to be developed, augmented and applied (Muir Wood and Fleming 1981, Reeve *et al.* in press) to many other maritime projects since this time.

Interactive models for the civil engineer of the greatest complexity, classified within the area of Computational Fluid dynamics (CFD), concern reactive processes. For example, models of pollution of water bodies may relate water flow to effluent discharges, with interactive chemical and biological reactions occurring. Comparable complexity is found in modelling fires, with air flow, including turbulent diffusion, heat generation and transmission processes all related to the temperature-dependent chemical processes. These are models of such complexity that the engineer is tested to the full, simply in the selection of the features of principal importance and their interactions.

It is important that the engineer entrusted with overseeing numerical modelling should, during his career, have been adequately exposed to the design and construction of prototypes. The virtual world of three-dimensional visualisation of the idealised model, seen from a variety of viewpoints, too easily obscures the problems of scale, of the circumstances in the field, of the intermediate phases of construction which may be the most vulnerable. Too often, the operation of the numerical model is undertaken by those lacking insight or practical experience of the problem to be solved, even to a degree of inability adequately to communicate with those interested in the results. This is a circumstance of potential, and increasingly actual, risk not often foreseen.

5.6 Design for the future

Civil engineering had once a distinct definition as the counterpart of military engineering. For many years, subsequent to around 1830, civil engineering seemed to settle comfortably into the role of engineering construction and maintenance of infrastructure (the word had yet to cross the Channel from France) works for civil and military purposes. Initially, these works responded to clear, specific demands to which there was no coherent, articulate opposition. From around 1950 there were hints of change as concerns began to be expressed on the social and environmental consequences of the civil engineer's major works. At the macro-scale, would a large dam affect the local climate? At the micro-scale, was there a more sensitive way of routing a new road? This was a new experience for civil engineers who had hitherto believed their works were benign, certainly causing negligible detriment by contrast with the effluvia from the 'dark, satanic mills' of manufacture, from the earliest days of the industrial revolution.

The attention to systems by the civil engineer dates from the appreciation of the need for holistic design to reconcile utility, economics, and social and environmental issues. Since the 1970s, the elusive quality of sustainability has added a new dimension to the qualitative factors. Communication between those engaged in the design process, as defined in Section 5.1, needs in consequence to extend well beyond the traditional major sub-disciplines (e.g. structures, geotechnics, materials and hydraulics) of the civil engineer's curriculum. As the overall complexity increases in each of these sub-disciplines, as well as in the social and environmental issues, voices are raised to suggest that the civil engineer of the future may no longer universally need the 'hard centre' of the traditional elements based on mathematics and physics. This notion loses sight of the plot. The civil engineer's function is most effectively to apply technology to the benefit of his Client, not forgetting the wider interests of society and the care of the natural world. It would be strange indeed if the engineer were to be expected to have no more than a qualitative, descriptive knowledge of the science behind engineering concepts. The boundaries between the several different engineering disciplines will continue to break down, with engineering becoming an increasingly continuous spectrum of disciplines with overlapping institutional allegiances. The first requirement of ability to communicate across the engineering disciplines will come through the commonly shared physics. The mathematics of their derivations will be found to be a link between quite different phenomena. The engineer is no longer protected from wider concerns by having his responsibilities confined to one familiar discipline. As described in Chapter 7, a vital feature that will distinguish the engineer is the ability to apply science through technology by means of systems working. Applied mathematics and physics will continue to hold the key to this ability. Fewer engineers will maintain familiarity with the latest developments of any one engineering sub-discipline. The ability to communicate appropriately, possibly through intermediaries, with such specialists – who will remain indispensable – will be an essential element of the design system. The engineer's education will, therefore, need to respect the three degrees of awareness, competence and expertise of Fig. 5.3, described in Section 5.1. The expertise represents depth required of the specialist, competence, the combined breadth and depth of the generalist, awareness, the ability to communicate with those others who will contribute to the rounded quality of the engineer's work.

5.7 The role of the Proof Engineer

The concept of *Prüfungsingenieur*, originating in Germany, was imported to Britain as 'Design Checker' (later more commonly described as Proof Engineer) for major bridge structures following problems with box girder bridges. This was at a period when numerical modelling was being introduced and there was a real concern for the reliability of the numerical models. There should also have been concern as to whether the models were being required to answer the critical problems, but this is another issue. It is not coincidental that a cluster of problems of bridge collapses occurred between 1969 and 1971 (Fourth Danube Bridge, Vienna, November 1969; Milford Haven Bridge, Wales, June 1970; West Gate Bridge, Melbourne, Australia, October 1970; Koblenz Bridge, Germany, November 1971). At this time, recent graduates in structural design had been trained in the use of numerical models, allowing, for example, more comprehensive study of stress distribution in plated structures. Meanwhile, the older engineers continued to use traditional analytical approaches. The most senior engineers had great experience not only of the critical features of structural design but also of the questions to be posed that affect the complex interfaces between design and construction. From my experience of the period, I believe

Systems and design

that problems in communication at a time of increasing structural complexity contributed to these failures.

I fully accept the desirability for external checking of a complex structure, where the checker has access to all design data of acceptable reliability: material properties; loading; dynamics; wind; temperature; and seismicity. I do find considerable problems when the Proof Engineer is introduced into areas in which the design process is less well defined. In the 1970s, for example, I was invited to advise Government on the intention to extend the function of Design Checker to tunnelling. I emphasised in my report that the design process must then include site investigation and its interpretation, including the discarded evidence and hypotheses, the application to the chosen technique of tunnel construction and the expected quality of control, including any departures discovered during construction from the design assumptions. These are all factors to contribute to the assessment of a satisfactory design overall. Unless, therefore, the Design Checker was present and engaged throughout all such activities and decision-making, acting as a 'shadow designer', his testimony could do no more than confirm that the process of analysis was acceptable for the stated assumptions. I concluded, therefore, that the degree of assurance to be provided by a Proof Engineer for tunnelling, confined to the design analysis, was illusory. It might even discourage innovation. I recommended that money would be better spent on employing a highly competent designer in the first instance.

Experience since that time only confirms my viewpoint. Proof Engineering confined to checking the numerical analysis of the design, but removed from other vital aspects, encourages the checking to be undertaken by structural analysts who may be engineers without experience in the appropriate field of application of the design. Two factors appear to have encouraged the increased dependence on Proof Engineering:

(a) The increasing complexity of the design analysis, for which perhaps it is both cause and effect.
(b) The increasing practice for the project designer to be employed by the Contractor, arousing suspicions by other interested Parties that safety might be jeopardised by cutting corners.

Proof Engineers are often appointed on price competition. It is unlikely that the tendering consultant, competing against others of lesser experience, will enlarge his scope of work beyond that specified for the function. This is an obvious instance in which the Client would benefit from allowing the consultant, here the Proof Engineer,

Civil engineering in context

to agree his own terms of reference. I believe that the Proof Engineer would be better employed under terms of reference requiring the following steps:

(i) Establish criteria for the acceptability of the design.
(ii) Study summary of the site investigation procedure and interpretation of results, ensuring revelation of any discarded 'rogue' data.
(iii) Study the nature of construction risks affected by design, including possible effects on third parties. Procedures for their control or elimination to be established.
(iv) Undertake simple (not numerical) analyses of failure modes and mechanisms, with sensitivity analysis, in order to pose particular questions on these issues to the designer.
(v) Taking a broad view of possible problems, including a possible comparative review recommended for the engineering geologist and other engaged specialists, to enquire whether each problem has been adequately addressed or whether, on the contrary, sound reasoning leads to its elimination.
(vi) Establish the degree of control by the designer on the construction procedures to ensure that these satisfy assumptions made in the design.

Alternatively, the function of the Proof Engineer might be achieved through a process of catechism, largely by setting explicit questions on the lines above to the designer. Familiarity with a project may encourage blindness to a particular risk by those most directly involved. However, the Proof Engineer must put himself mentally into the problem area in order to understand the potentially critical issues. The German word *gestalt* (form) has been broadened internationally to equate to something close to 'holistic'. I am advocating that both the designer and the Proof Engineer should be expected to exercise such an approach in preference to the narrow role of checking the structural analysis, which is much more likely to be correct than its assumptions, its specified objectives and interpretation. The Employer who engages a Proof Engineer might then expect to gain value for money. He should not expect such a valuable service from one who has been engaged at least cost for a function of uncertain and inadequate content. The essence of the matter is that the Proof Engineer should not be undertaking a narrow function divorced from the total system. There needs to be consideration of how practically to combine the full function of Proof Engineer with the safety features of CDM, while maintaining continuity in overall responsibility.

6
Learning from experience

6.1 First impressions

I have not set out to write an autobiography. I readily accept, however, that the opinions and prejudices, especially the prejudices, expressed in this book must be affected considerably by life's experiences. My object in this chapter is to recall episodes, memorable at least to me, some apparently trivial, but then perhaps there is some truth in David Copperfield's observation that 'trifles make the sum of life'. These episodes have also influenced my chosen profession and, in so far as this lies within personal choice, career. Some of the incidents carry warnings for others who find themselves in comparable circumstances. Much of the learning process of an engineer concerns the absorption of the broader significance of each experience. One question needs to be answered before it may be sub-consciously posed. Why did I choose civil engineering? Immediately, I remark that, while there are many attractions to other options, I do not regret my choice. I have, from my earliest memories, enjoyed trying to understand and solve practical problems. I suspected from what I had seen of the times, prior to making a conscious decision, that there were better ways of undertaking engineering works. University provided a framework for logical approaches to unfamiliar problems. I also thought, and continue to think, that I had a social conscience. The satisfaction of these considerations was, I guessed, as likely to be found in civil engineering as elsewhere. And finally, I have always enjoyed the open air and did not see myself as following the long but declining farming tradition in the family.

Those of us who have worked in the construction sector continuously since the Second World War can claim to have seen interesting times. The war revealed how obsolete were much of our traditional practices and equipment in many branches of engineering. The ensuing period of deprivation and peace allowed a similar reflection, but in a broader sense, as we surveyed our infrastructure (although this was a term yet to cross the Channel). Generally, engineering had fossilised. Dams and bridges were subjected to ridiculous architectural embellishments. Only the most *avant-garde* architects paid the slightest attention to the opportunities of adventurous engineering.

I need to go back a little further to indicate the formative features of my own perspective. My father was an Admiralty civil servant. Apart from a period in Malta, between the ages of around three and six (where I gained the impression that the Mediterranean Fleet was there to provide ingenious children's parties – with balloons fired by compressed air from the main armament, capstans serving as roundabouts, overhead wire rides from the masthead, ascent by climbing nets, all activities to be deplored by a present-day Health and Safety Executive (HSE)), later my brother and I lived for three years in Chatham's Naval Dockyard. We were literally in the dockyard, at No. 8 The Terrace, now a spectacle from the 'Historic Dockyard'. We had freedom to roam, into the machine and heavy forging shops, into the ropery and around the graving docks and fitting-out basins. Here, the scale and continuity from the nineteenth century were palpable, accompanied by the sense that the activities were expected to carry on indefinitely, in the support of an Empire, in much the same style.

We enjoyed periodical entertainment when a timber-importing vessel unloaded in an impounded basin. The timber was destined to be floated around to the mast pond. Each massive baulk of timber was launched athwartship into the basin by a sling lifting the quayside end. The dynamics are illustrated by Fig. 6.1, the baulk diving obliquely into the basin. The intention was for emergence towards the side of the vessel, for capture by the raftmen. Occasionally, however, a baulk would escape, emerging away from the vessel and streaking across the basin like a torpedo. As illustrated by Fig. 6.1, a couple resulting from the weight of the baulk and the buoyancy force is opposed by drag transverse to the baulk, while re-emergence in the direction of dive is mainly opposed by drag along the baulk. The motion is then determined by rotational and translational accelerations. As with so many problems of hydrodynamics, the controlling forces are readily identified but the balance between them, the consequence unpredictable for an individual instance. We did not then know that we were observing a phenomenon now classified as a 'bifurcation', a sub-species of the theory of 'chaos'.

Learning from experience

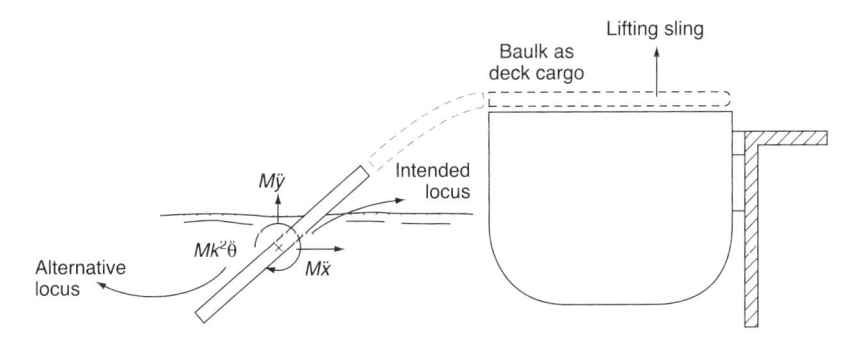

Fig. 6.1 Off-loading of timber baulks at Chatham

These years were certainly impressionable, but why did nobody ever mention the surviving traces in the dockyard of the work of Sir M. I. Brunel, his sawmills, by that time a laundry, and connecting tunnel to the mastpond?

6.2 University – the foundation for curiosity

The influences from Cambridge are many and various. No doubt a number of the more lively lecturers had already been called up for the war. The lectures felt timeless, not for their duration but for the impression that the pattern and content had survived the years. Supervisions were sound and thorough but not adventurous. As an example, the text book on electricity was the *Naval Electrical Manual* of 1928, which had superseded, but not too comprehensively, the previous edition of 1911. The electron receives mention but not electronics. Our course was rightly termed Mechanical Sciences, not Engineering. We learned about well-established technologies with their theoretical backgrounds, a course largely of applied logic in Newtonian mechanics. Aspects of engineering which were later to become of central interest to me, geotechnics and maritime engineering, were scarcely touched, beyond Rankine's earth pressures and classical long-wave theory. Since the art of engineering has to be learned from exposure to practical issues of a constantly changing nature, the most important aspect of a university course must be that of teaching the student how to learn, and to apply techniques of exploration of phenomena and of the 'what if' of innovation. The most memorable lectures to this day were from H. A. Webb on mathematics, who used as illustration examples from his experience in the First World War, including the optimisation of the design of biplane struts and the calculus of variation (rather advanced, I am told, for engineers!). The greatest problem in teaching mathematics is that of motivation. Those who have experienced difficulties of this nature should read Webb and Ashwell (1959). The feature that most impresses engineers is that of utility. From the earliest, mathematics should be taught to emphasise the use of each step to explain a practical issue; the remainder is for the pure mathematicians, at least until it too finds an application.

A debt I will always owe to the Tripos course is that we were taught to aim towards simplicity but to question every step – and we did, on one occasion to the embarrassment of a lecturer who had integrated, in a lecture of obvious vintage, a variable as a constant. (Some years later it was darkly hinted that pragmatism without underlying theory was creeping into the syllabus!) The Inglis Bridge, an unsuccessful rival to the Bailey Bridge, was the most evident sign of research activity. Charles Inglis, Head of Department, who gave occasional lectures on disparate subjects, sums up the philosophy of such a course neatly in his Presidential Address to the ICE: 'The spirit of education is that habit of mind which remains when a student has completely forgotten everything he has ever been taught' (Inglis 1943), not I subsequently learned disappointingly an original observation.

A vital feature of a successful university course must be that of motivation. It is in consequence valueless to argue if one approach is better than another, without considering the features which engage the interest of the student. High quality, the course must be; but motivation depends also on the type of student. There is in consequence great virtue in a diversity of courses. A strength of our profession lies in the diversity of routes followed by its practitioners; a weakness lies in the inadequate mutual respect between the theory and practice. Over-precise prescription of an engineer's profile must be avoided – this is a lesson for the Engineering Council and the Joint Board of Moderators (JBM) of the Institutions of Civil Engineers and Structural Engineers.

As an aside, in 1941 I worked for the summer vacation at Vickers at Weybridge, assembling Wellington bombers. I was amazed by the low intermittent rate of activity demanded for low-skilled work by skilled machinists and fitters – one had to make lighters out of sheer boredom – and this was at a most desperate period of the war. By contrast, harvesting in Kent during the previous long summer as the Battle of Britain was fought overhead was the most enjoyable back-breaking work!

6.3 The Navy at war

While at university there was a satisfactory basic approach to engineering problems, and the problems were far more evident than the opportunities for applying solutions. We were reminded that, as graduates of a general engineering course, we would, initially at least, be just as useless in any branch of the subject. The Royal Navy set about both amelioration and emphasis of this humility. The first was effected by an immediate commission, the second by spending nine months in working through the shops of Thornycroft's shipyard at Woolston, Southampton. Following my impressions from living in Chatham Dockyard, the view was confirmed that shipyard practice had scarcely changed in 50 years, that total reliance was placed upon the skills of the craftsmen, the welders, the boiler-makers, the fitters, the founders, the braziers and the pattern-makers in particular. By contrast there was little evidence of engineering direction and none of concern for innovation. This episode is a vignette of the wider scene of a stasis in technology, compensated for Britain by access to secure commonwealth and colonial markets which for a long time had obscured, at least to the complacent, the obvious consequences in terms of international competitiveness. The lessons for today remain obvious.

The technology, largely based on the skills of craftsmen, had changed so little since the beginning of the twentieth century. The ships were riveted, all welding was manual, steam pressures had increased but the Parsons steam turbine remained the power unit. The management showed remarkably little interest in the technical aspects of the work, their remoteness doubtless encouraging the strict demarcations of work between union factions. Since I moved from shop to shop, it was difficult, as a fitter, for example, to remember that I was no longer allowed to drill holes – this was the prerogative of the boiler-makers. I received stern warnings. I can only assume, from the invisibility of management, that work was reimbursed cost-plus. The skill of many tradesmen was impressive, not only in their particular craft but also in the organisation of their work. The aristocrats were obviously the pattern-makers, working entirely in wood, designing patterns for intricate castings, capable of assembly, dismantling and removal without damage to the mould. This required considerable ingenuity and spatial visualisation.

A curious experience occurred as I awaited in Alexandria for my first sea-going commission, living in the cockroach-infested base of a lighthouse on Ras-el-Tin, not too far from the site of the ancient Pharos. Italy had surrendered and two vessels, the battleship *Vittorio Veneto*

Civil engineering in context

and another, sailed under escort into the harbour. I was detailed off, as supernumerary and possibly expendable, to deliver a signal to each captain. To my consternation, and in contrast to the normal lot of a very junior, inexperienced officer, I was piped aboard with great respect. I was offered, and accepted, a tour of the ship, immediately aware that I was among reluctant enemies who belonged to a remarkably friendly nation.

My first ship, HMS *Carlisle*, being dive-bombed my third day at sea, I was posted, via a brief stay aboard HMS *Dido* to a destroyer, HMS *Petard*, to work towards my watch-keeping certificate. I was possibly the first engineer officer to do so in this manner, since watch-keeping engineers only served aboard larger ships. So, I was set to undertake successively the duties of the engine-room crew, starting in the boiler room in the role of Stoker second class, finally alongside the Chief Engine Room Artificer (ERA) in charge of the watch. This experience once again contributed to forming a perspective from an unfamiliar vantage point. A ship at war provides a good opportunity to reflect on the interdependence for survival on the conscientious performance of particular skills, an example of symbiosis. With the technology so obviously traditional, familiar and mature, innovation would require a new regard for the system. There was, therefore, little spur towards technical curiosity.

Before leaving the Mediterranean, we were granted a week's leave from Haifa. The navy would ferry us to any place of our choice in Palestine, thereafter we would be on our own. The popular choice was for the flesh-pots of Tel Aviv. Two of us (my companion was later Chief Education Officer of Sheffield) determined to make our way to Bierut, a somewhat ambitious venture since we possessed only a 'political' map at a scale of about 1:500 000, no compass or walking gear. Moreover, I (having left a foundered ship) had only the remains of a fuel-oil-soaked naval identity card lacking photograph. We were taken to the shore of the Sea of Galilee (Lake Tiberius) whence we made our own way to the foothills of Mount Hermon, from which a spring gushes from a 'portico' with pediment, carved into the rock, Petra on a small scale. We depended on finding shelter each night. The first was spent with the head man of the village of Baniyas, where we enjoyed a communal meal as the livestock were herded into another part of the building. We found our host was a politician, so, as we consumed our arrak, we discussed politics in our best French. At dawn we set off for our next destination, Mej Ayum, observing the flourishing border trade in smuggling, as we set off along the

Learning from experience

slopes of the Jebel Libnan of Lebanon. We were briefly arrested, there being some sensitivity towards the Vichy French; my companion's black beard apparently attracting accusation of a cleric in disguise. We were taken to a British Army field centre and soon released on our own recognisances, with a caution of the danger of the many lethal insects that might beset us, demonstrated by examples in bottles of formalin. Our next problem was the encounter with the River Litani (many years later the site of major problems of a hydro-electric project). Our map showed no bridge, we were in an uninhabited district and the odds appeared even against finding a crossing in either direction. We chose right and found a crude tree-trunk bridge. We stopped for further nights at Jezzin and Beit-el-Din where, on each occasion, we stayed with a poor household glad for the small contribution from our company. One, who appeared to survive on remittance from America, seemed to subsist on a sort of porridge, scooped from the same container for each meal. We could buy eggs, fruit and vegetables close by.

We passed through several villages, Muslim, Maronite and Christian which appeared to live together peacefully, without concern for historical conflict. It is an irony that in wartime I visited several pacific countries, and the Lebanon, Palestine, Sri Lanka and other parts of the British colonial empire, today continuing to pay the price of racial and tribal antagonisms whipped up by politicians.

We reached the Damascus–Beirut road and thence found our way to Beirut. Here, after several days growth of beard and limited washing, we were rejected as too scruffy by the Officers' Club. The NCOs' Club was more accommodating. We spent a comfortable night before a bus journey down the coast road passing (regrettably) Sidon, Tyre and Acre on our way.

In due course, the ship's Chief Engineer was relieved by E. W. K. Walton, who had experience of the Antarctic, was later a schoolmaster at Malvern School and Member of Council of the ICE. He initiated the scheme 'Opening Windows on Engineering' taking young engineers into schools to explain aspects of engineering; I helped him to obtain support from the institutions for this pioneering activity. To return however to *HMS Petard*, this commission was spent largely in convoy duties between Aden, Bombay and Colombo. For every ship, there is a damage control diagram indicating the effects on trim and heel of flooding of a compartment. I was curious to explore how stability might be affected as the result of such flooding, through the combined effects of list and loss of metacentric height. I prepared a simple diagram without thought of practical application. Only a few days later, on

12 February 1944 (I am reminded of the date by a letter in *The Times* by Mr Guy Yeoman in February 1994) an application occurred. A large Japanese submarine, I-27, was brought to the surface by depth-charges after sinking a troopship, *Khedive Ismail*. Our sister destroyer, *HMS Paladin*, essayed to ram the vessel but, being ordered to desist, attempted a violent hard-to-port, hard-to-starboard manoeuvre, but not sufficiently rapidly to prevent tearing off a length of her hull plating against a hydroplane. I could estimate the compartments affected, consulted my diagram and predicted that, in a calm sea, she would not capsize. We towed *Paladin* to a familiar coral island, Addu Atoll, whose shallows were a tropical aquarium, while a patch was installed and then to dock for a new bows section.

When later I encountered the US Navy, and explored their machinery and layout, my concern for British complacency in ship design between the wars was further enhanced. We had failed to modernise, a view from the engine-room which complements Churchill's 'view from the bridge' of *The Gathering Storm*.

While at sea, the crew, most of whom were HO (Hostilities Only) and many well educated, organised a periodical question-and-answer session, broadcast over the ship's public address system. In due course, I was invited as the guest officer. I enjoyed the occasion for a free expression of opinion and discussion until the question arose 'Had the peacetime services adequately prepared for the war'. While it was probably unfair to put too much blame on the services rather than the politicians, I nevertheless launched into the subject, with many personal observations to support my criticisms. As a summary, without too much thought for its declaration in more temperate language, I declared that the peacetime services had been largely parasitic on the nation. I thought no more of this until I reached the wardroom, to find the Commanding Officer and the First Lieutenant, the only regular officers in the complement, apoplectic with indignation. A spirited debate followed with support otherwise for my point of view, if not for the precise expression. Shortly afterwards, I was introduced to a very senior naval officer by name and rank, followed by 'he thinks I'm a parasite', which appeared to be received with equanimity without the need for further explanation!

After a brief period aboard *HMS Suffolk*, I spent a period on the Tyne where Liberty ships were being fitted out to serve as aircraft repair vessels for the Pacific Fleet Train. My first recollection is that of advising that setting out, aboard a ship afloat, of an aircraft engine test rig should not be based on use of a spirit-level. My final commission

Learning from experience

at sea was aboard HMS *Newcastle*, serving out my time after a period at sea by supervising a refit. In 1980, I had the opportunity to join a visit by the Royal Academy of Engineering to the new Type 42 Frigate *Newcastle* at Portsmouth. The comparison with her predecessor was remarkable, not least with the degree of technical understanding of a new generation of equipment by the senior artificers. I was also struck by the degree of dependence on the integrity of the central nerve cell of the vessel, quite unlike the scope for improvisation of 40 years previously. When a few months later the Falklands war occurred, I recalled the cheerful relations with the Argentinian, apparently congruent, frigate alongside.

6.4 A start at civil engineering

When (my) war came to end in 1946, my wife and I were living at well below today's definition of poverty level (or perhaps not, on reflection, since the general standard of wealth was low). She resigned from the Royal Naval Scientific Service to start a family. She and I, designated after four years in professional engineering as a draughtsman by the Southern Railway (soon to become British Railways – Southern Region), learned how to survive on £6.25/week (which included a special London allowance of £1.25!). There was scarcely any activity in the industry, apart from the first measures to recover seven years deficit in maintenance of roads and railways, so there was no luxury of choice. It also deserves note that the ICE had made no provision for any form of membership to help those caught up in the war. I was prepared to wait to demonstrate that four years experience as a marine engineer might be of value to a career in another branch of engineering. I continue to feel resentment, however, that, doubtless with many others, I was cut off from the profession at a time when help was most severely needed. I subscribed to the Proceedings and irregularly attended occasional meetings. In due course I qualified at the age of 28.

I spent two years in a Divisional Engineer's office, predominantly concerned with the alignment of track, especially transition curves. This was a simple process comprising a survey, by measuring versines, a manual process of relaxation to determine 'slews', the establishment of monuments to record the new line and cant (super-elevation). I soon realised that the calculation could readily be undertaken during the associated train journey, during the considerable period spent in traversing the area of the South-West Division. The other main

Civil engineering in context

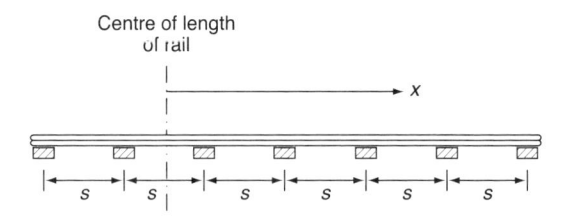

Fig. 6.2 Reference diagram for temperature effects on long rails

maintenance operation concerned the blanketing of track to prevent clay 'pumping', a phenomenon studied by A. H. Toms, the undervalued, highly experienced, geotechnical researcher for the Southern Railway. Weekend possessions for each length were planned as military operations, by those recently released from the armed services. On a Monday morning, a meticulous review followed, with an overrun, even of a few minutes, treated as a tactical defeat. 'What a falling off was there' in such standards since that time.

The first long rails were being introduced experimentally at Wimbledon. During the hot summer of 1947, the six or so members of the permanent way office undertook measurements of temperature and rail extensions throughout the day, a life-threatening operation on a busy four-track area of poor approach visibility. The problem may be formulated in the following manner, referring to Fig. 6.2.

A pair of long straight rails of length $2L$, with total area A, is subjected to mean temperature rise, T, uniformly distributed along the rail, whose temperature coefficient of expansion is α. If Young's modulus for the rails is E and sleepers at spacing, s, resist longitudinal extension of the rail, u, by resistance Bu, and the stress in the rail is represented as f, the circumstances may be represented by two equations:

$$\frac{\partial u}{\partial x} \equiv \frac{f}{E} + \alpha T \quad \text{and} \quad A\frac{\partial f}{\partial x} = \frac{Bu}{s} = Ku \text{ say, where } K = B/s \quad (6.1)$$

where x is measured from the centre of length of the rail.

We know that at $x = 0$, $\partial u/\partial x = 0$. And at $x = L$, $f = 0$. Equations (6.1) provide relationships between f, u and x solved as:

$$f = -E\alpha T\left(1 - \frac{\cosh Dx}{\cosh DL}\right) \quad (6.2)$$

and

$$u = \frac{\alpha T \sinh Dx}{D \cosh DL} \quad (6.3)$$

where $D^2 = K/EA$, so that, at $x = L$:

$$u = \frac{\alpha T \tanh DL}{D} \qquad (6.4)$$

which, when DL is large, tends towards:

$$u = \frac{\alpha T}{D} \qquad (6.5)$$

Temperature movements at rail expansion joints were found to correspond to the free movement of a length of rail of about 20 m, by recollection, thus providing a value for B for these particular conditions. The experimental data provided curves which adequately fitted the simple hypothesis of linear relationship between rail movement and sleeper resistance. As might be expected intuitively, the corresponding length of 'free' rail is unaffected by the value of temperature rise. Hence, the remainder of the straight rail, of whatever length, needs to withstand the full temperature-induced compressive stress.

During a short period in the bridge office, I surveyed many iron and masonry bridges, endeavouring to introduce a more systematic filing of inspector's reports, as a result of finding great difficulty in assessing trends of movement. I designed a reinforced concrete lining to an overbridge for the Duke of Northumberland's River to Hampton Court – never, I think, built – where punters tended to penetrate the clay puddle bed and cause water to spurt from the brick piers into the cab of the motormen. This provided the occasion to estimate a free water surface by a Southwell form of relaxation, an approach which, in due course, evidently satisfied the Institution's examiners. At that time, a Bill of Quantities supported by taking-off sheets was also obligatory. The railway gave me opportunity to prepare these for a scheme of surfacing and drainage of a goods-yard, provided I worked in duodecimals, the odd predilection of Quantity Surveyors. Why, in consequence, did they not invent symbols for '11' and '12' which might have made the practice defensible, if bizarre, prior to the conversion to SI units?

One became an authority on railway canteens during this time of food shortage and rationing. During a spell in setting out a complex junction nearby, I learned the superiority of the provisions made at the Nine Elms' locomotive depot for the industrious artisans. However, a newspaper was needed to serve for protection from the greasy bench, cutlery required a deposit and, for those who needed to stir their tea, the spoon was chained to the counter.

6.5 Folkestone Warren

Renewal of movement of coastal landslides at Folkestone Warren was once again threatening the main railway line between Folkestone and Dover. This was the only sizeable railway project in the south of England and I was fortunate to be offered the post of Resident Engineer for the remedial work. I was responsible to E. C. Reed, an engineer of the old school with considerable practical experience, kindly, avuncular, short-sighted and a keen photographer; he shared this latter enthusiasm with A. H. Toms, to whom also I was responsible for continuing the investigations which he had started a few years previously (Toms 1946), particularly related to estimating benefits to be expected from improved drainage and toe weighting. Folkestone Warren has inspired much study and reflection from engineers and geologists, a partial bibliography being provided by Trenter and Warren (1996).

Soon after my arrival in 1948, I realised that the copious macro-fossils of the Gault clay, including many varieties of Ammonite, could provide markers of geological age and could therefore help in reconstructing the sequence and extent of the numerous landslides, from cores recovered from boreholes. I believe that this was the first time that such evidence had been exploited by engineers. Learning how to identify the dominant 'marker fossils' under the expert guidance of Raymond Casey (later Fellow of the Royal Society) of the Geological Survey, I prepared a simple monograph, illustrating the principal marker fossils for use by others at Folkestone Warren. I passed a copy to the Geological Survey who asked me to attach a 'health warning' to any copy which might fall into the hands of a palaeontologist to avoid any imprecision in illustration leading to identification as a new sub-species. This palaeontological information was used at the time, identifying stratigraphy adequately to establish the expected improvements in stability.

The Chief Civil Engineer, V. A. M. Robertson (VAMR), accompanied usually by Sir William Halcrow and his fellow Partner, H. D. Morgan, travelled from time to time to the Warren, in the Chief Engineer's observation car which would be shunted to a siding alongside the Warren Halt – incidentally, right across the back of a slightly active slip. An excellent lunch was always served on these occasions in the dining-car. On one such occasion I was offered a ride back to London with the party. I witnessed, surely, one of the last examples of deference as lengthmen and gangers along the route raised their caps as we passed, acknowledged by VAMR. I never understood Halcrow's commission in relation to the Warren. From time to time, I saw sketches but always after we, at site, had designed the work in far simpler fashion and

Learning from experience

greater detail. Our particular problem lay in designing around building materials available immediately post-war. All the dowels which coupled the units of the block sea-wall, for example, were cut from hexagonal bars known to our excellent Inspector of Works to be lying at a depot in London, presumably originally obtained years previously as stock for turning nuts and bolts. The work of beach weighting, a block wall around three sides of chalk fill with a slab cover, was undertaken by direct labour locally recruited, largely working tidal shifts.

In 1994 I was invited to present a Paper on 'momentous occasions in geotechnical engineering' at the 13th International Conference on Soil Mechanics and Foundation Engineering at New Delhi. Not being allowed to plead that I had had no such occasions, I agreed that to me at least this first serious encounter with soil mechanics remained memorable and provided material for a considerably more extensive analysis (Muir Wood 1994). Even the most laid-back academic is unlikely to have delayed writing up his research for nearly 50 years. The periodical landslides probably started to occur at relatively frequent intervals since construction of Folkestone Harbour began to cut off longshore drift of sand and shingle from 1810. Landslides are recorded from 1765. From 1839, minor slides occurred around every ten years, with major slides every 20 years or so. Roman and later remains support the conclusion of little erosion over the years of the promontories at each end of East Wear Bay which lies off the Warren. My most recent analysis throws light on the geometry of the rotational part of each major slide and suggests a simple but possibly adequate approximation for the reactions between the rotational and translational elements of the slides – at least until a better means is proposed for modelling the high inhomogeneities of the phenomenon. Each landslide provided temporarily improved stability, until the combined effects of renewed erosion and high watertable caused the next movement. The timing would also be affected by the rate of weakening of the clay along potential shear surfaces, as a result of swelling that resulted from shear stress of the over-consolidated clay. A borehole of 1972 recorded the pore-water suction associated with this effect. Long-term stability therefore required considerable gain in safety factor by remedial works. I had previously (Muir Wood 1955) established direct correlation between water levels in the chalk, used as a source by Folkestone waterworks, and the rate of inflow to the Warren, deduced from yields from a spring in the adjacent Martello Tunnel, used for the water-crane at Folkestone Junction for steam locomotives.

Civil engineering in context

6.6 Another touch of research

Although I was told that I had a future career mapped out with British Rail, there was no encouraging immediate prospect at this time of low activity. I answered an advertisement, attended interview and assumed the post of Research Assistant to a new organisation to be set up by the nationalised Docks and Inland Waterways Executive at Hayes under Research Director, J. T. Evans. Evans had moved sideways from the position of Chief Engineer to the Severn and Trent Navigation Authority, perceiving the need for research ahead of the changing role for canals and the revitalisation – it was hoped – for the 'railway' docks. He had a lively and resourceful mind, and provided much interesting background from his experience. I recall, for example, his warning about the need for preparation for any encounter with the law. On an occasion of damages claimed against his Authority, he was defending the immaculate condition of a weir on the Trent, to be reminded that he had described it in an internal communication as a pile of rubble covered by a thin skin of concrete. As a sign of the times, we were considering the basic design of siphon controls for abandoned canals, many now restored to use.

This was an entertaining period, if not on a direct path for construction. I recall well a winter's day of snow and ice, travelling to and fro on narrow boats using Telford's Harecastle Tunnel to measure exhaust pollution levels. Far more lethal seemed the malodorous sulphuretted black liquid pouring into the deformed tunnel from disused mine workings. A longitudinal ventilation system was subsequently provided.

We were exposed to a variety of problems, many concerning defects and different forms of corrosion. We generally acted the part of facilitator or catalyst rather than researchers, first in understanding the basic causes and, occasionally, from direct knowledge, providing a solution. Our primary purpose was to undertake hydraulics research. We needed first to design a laboratory. Having no other technical staff to call upon, I prepared the drawings, after studying details of a few, more ambitious, hydraulics laboratories in Britain and Europe, particularly those related to internal waterways. To facilitate the grant of planning application, I added a few trees and other deft touches to give an air of architectural influence. We then needed a towing tank to study the drag characteristics of vessels, predominantly of commercial craft on inland waterways. Fortune had decreed that a tank with towing carriage of just the right size (perhaps we designed our own laboratory to this end?) figured in the 'Britain can make it' pavilion of the 1951 Exhibition on London's South Bank. At the auction after the end of the exhibition, I found myself bidding for the tank only against scrap-dealers, coming away in consequence with a bargain. We set up an information system. Although I cannot claim, during my time, that it made great impact on the engineers and others it served, it provided considerable opportunity to absorb current trends of technical and social significance to the industry from journals, also to meet a wide variety of people. Impatience did not allow me to wait to see the laboratory in business.

6.7 In consulting practice

In describing aspects of selected projects below, I recall that I invariably made a conscious effort to identify areas of uncertainty and hence of potential risk. A conscientious engineer does not wish to be associated with a project in which the Contractor is expected to bear the cost of encountering a problem unperceived by the Engineer. I believe that I have acted fairly in such respects and that this has served as financial benefit to my Clients.

6.7.1 *Tunnelling*

I joined Halcrow in 1952, responding first to an advertisement and later to a conversation with a Partner, V. A. M. Robertson, familiar from my railway period. I was immediately plunged into the design of watertight concrete tunnels for underground telephone exchanges in Birmingham and Manchester, followed by Oban, for the Atlantic cable shore terminal, and initial studies elsewhere. These tunnels were in in-situ concrete with waterproofing additives, cast in fairly short lengths with elaborate construction joints, incorporating a gland between outer sealing strip and inner mastic jointing compound. It was likely that, initially, the gland would tend to be blocked by calcite; it was therefore designed to be cleared by the use of rotating hammers on long shrouded cable, familiar to me from boiler-tube cleaning procedures. These were early days of PVC seals. I satisfied myself, using two squash-racquet presses and an electric fire, that welding by surface heating would make effective transverse PVC junctions. There were many other lesser operations. H. D. Morgan, Partner for these operations, had good technical judgement and considerable ingenuity, entertaining if irascible, but not always a reliable judge of character. I recall one such enthusiast, let us call him Dixon, who insisted that his gap-graded concrete mix was ideal for waterproofing purposes. Horace

Civil engineering in context

Morgan gave him leave to test his proposals for a pumped mix at Manchester. The Contractor objected that it would stick in the pipeline. Dixon averred that he would eat the mix if it did; he was already on his way back to London, however, by the time this occurred. We needed many fastenings to the concrete linings. Here again, Dixon had the answer with bolts set in mortar with applied electrical potential by means of a copper wire coil around the bolt, so that chemical salt deposition would occur at the bolt as anode. I attended the pull-out trials. The control, simply mortared in place, showed ample resistance. The experimental bolt came away easily; Dixon had overlooked the effect of gas bubbles in reducing the area of contact.

We were approached by the railway to supervise the construction of three new double-track tunnels near Potters Bar on the East Coast main line. A set of contract drawings had already been prepared in in-situ reinforced concrete. Horace Morgan accepted the appointment provided the tunnels were redesigned to make them capable of economic construction. This is by no means the only example I have seen of tunnels designed by structural engineers with no concept of how they could be built. It was my task to redesign the tunnels at high speed to honour the required completion date, the first lining in Britain to be expanded by jacks. In retrospect, this was a fairly over-robust design but several factors provide justification. Steam locomotives required provision of smoke shafts, with ample throats. There were also frequent refuges to be constructed as recesses. Limited cover near the portals would require a stiff (or alternatively reinforced) lining. For a shield-driven tunnel there is great merit in avoiding the need for subsequent excavation affecting lining stability. By means of cylindrical shear-keys between segments at openings, subsequently filled by earth-dry rammed concrete, and other devices, no internal temporary support was needed. In today's climate of obsession with the requirements for health and safety – well justified in principle if not in some of the risk-aversion detail – it merits note that as work approached completion, the railway authority realised that they had not obtained authorisation for the specified internal profile, which encroached on the standard structure gauge. A wooden template of the rolling stock kinematic envelope was provided for a visit of the Railway Inspectorate who accepted the arrangement, which was as well!

The experience at Potters Bar (Terris and Morgan 1961) led to a new approach to expanded linings for the Victoria Line for London's Underground. An enlightened engineering management had decided that the construction should be preceded by an experimental tunnel where different forms of lining could be tested for performance and buildability. A colleague, Jack Donovan, had already designed the Donseg lining of alternately tapered segments. Such a lining for the then current tunnel ring widths was only practically suited to tunnels less than about 4 m diameter. I set about designing a lining based on the concept of shaped convex-to-convex joints, to permit a degree of deformation while maintaining the force towards the middle of each segment. At the same time, a theory was developed on the sharing of load between the ground and the tunnel lining, used, with adaptation, to the present day – and probably more reliable than two-dimensional numerical methods applied to the three-dimensional state of stress distribution close to the tunnel face. It was obvious that considerable savings in cost could be achieved through such flexible linings, also that settlement could be better controlled by such linings, expanded against the ground, than by the use of the traditional bolted and grouted linings. This was demonstrated most forcibly later for the Cargo Tunnel at Heathrow (Muir Wood and Gibb 1971) which achieved a ground loss of 0.3% (ratio of surface settlement to excavated volume), by contrast with 1–2% expected for other forms of construction current at that time.

The next major challenge arose from the Clyde Tunnel, twin road tunnels to be constructed in a variety of jointed Carboniferous rocks, glacial deposits and alluvium. The available technology determined the use of compressed air and the use of cast iron linings, which needed to be 'stiff' for the weakest ground, to avoid distortion of the ring. The Parliamentary Act under which the tunnel was to be built decreed an inadequate size and a cramped layout for the initial approaches. Ventilation requirements, even for a tunnel of modest length with the exhaust gas outputs of the time, were demanding. There was also the need for a cyclist and pedestrian subway beneath the road, to meet the needs of the shipyard workers. The solution was to situate supplementary fans in mid tunnel, drawing fresh air through the subway. The drawback of this expedient was that the cyclist would find the breeze always downhill, on a gradient of 6.25%. An episode during construction deserves mention beyond the published account (Morgan *et al.* 1965). The tunnel was to replace a vehicular ferry and, at the North shore, passed directly beneath the ferry berth. Also at this point, ground cover would be small, predominantly in sands and silts. Compressed air tends to escape up against the face of a solid structure in a soil, akin to the phenomenon of piping, but examination of the as-constructed drawings for the jetty, prepared by the

Civil engineering in context

contractor, indicated the toe of the timber piles as being 2 m or so clear of the tunnel. Here I made two errors; first, in concluding that a contractor paid by length of pile would not underestimate it – he did by a metre or so; second by failing to appreciate the likely loosening of the piles by years of buffeting by the ferry as it docked. The consequence was a 'blow', fortunately no more than a minor delay and cost, for a project built within time and budget. Tunnel drives were completed in compressed air by driving the shield into each portal block backfilled with clay. The broad bifurcation of the tunnel approaches, in preparation for improved access, required the design of large diameter concrete tension piles with enlarged bases, achieved by a simple analytical design, adequate for the purpose. A curious complexity seems to have attended such problems elsewhere in later years.

The phenomenon of earth tides, caused by the slight distortion of the earth as a result of differential gravitational pull by the moon and sun has been known for many years, the effect revealed from continuous recording of slight variations of water level in wells. Volumetric strain of the earth's crust caused by the moon's semidiurnal tidal component is estimated (Melchior 1966) at around 2×10^{-8}. An unconfined aquifer will therefore not be associated with a measurable differential change in water level relative to the earth's surface. Following a major flooding incident on the Orange-Fish Tunnel described by H. J. Olivier (Olivier 1970), continuous variations of water levels in boreholes close to the tunnel were recorded over the New Year 1969–70, as illustrated by Fig. 6.3. Tides are caused predominantly by the tangential tractive component of the differential gravitational force, the effect being greatest at an angle of 45° to the line between the earth and the disturbing body, moon or sun.

Ocean tides are modified by factors such as topography, friction, resonance and Coriolis force. Fluctuations of earth tides may be expected to correspond directly to the theoretical combined effects of the several tidal components, designated as S_1, M_1 for the diurnal components, S_2, M_2 for the semi-diurnal components, apart from minor secondary effects caused by ocean tides. The dominant component is M_2. From the trace for borehole NL9 on Fig. 6.3, the double amplitude for the M_2 component is seen to be about 200 mm. The tunnel was situated at 31°S, 26°E and hence near to the plane of the ecliptic at this time of year. The phenomenon was found to be associated with a previously unsuspected extensive fault system trending approximately east–west, roughly in the plane of the ecliptic. The site investigation in several

Learning from experience

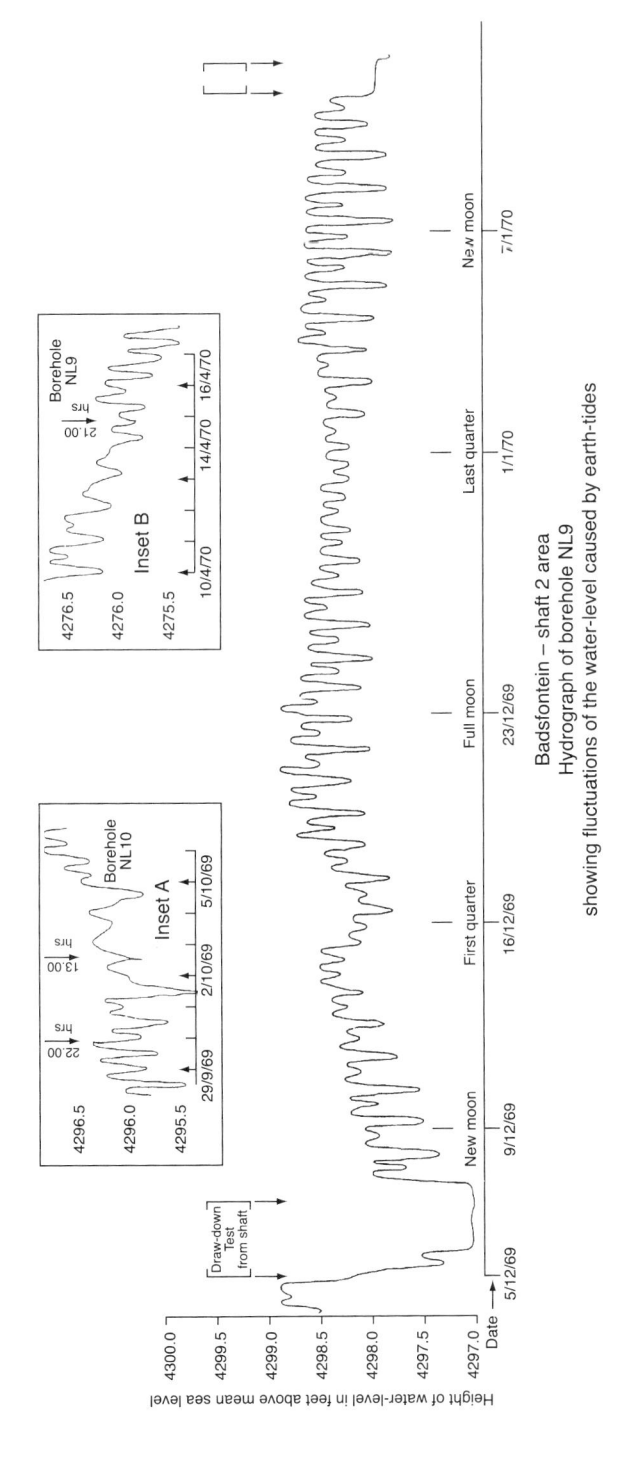

Fig. 6.3 *Earth tides – evidence from Orange-Fish Tunnel (After Olivier 1970)*

131

phases between 1947–67 had identified numerous faults associated with dolerite intrusions but this fault system was not of such nature. Olivier also records that seismic shocks during the period of recording the water levels affected the pattern of movement over several hours. The evidence points to strain locally concentrated at the fault system.

Surveying by theodolite relates all levels to the geoid, the surface of equal gravitational potential. The movement of the earth's crust as a consequence of earth tides would not therefore be a feature for concern. The introduction of Global Positioning Systems (GPS) requires earth tides to be considered for surveys of high precision on long base lines. Earth tides are, therefore, no longer just an interesting but irrelevant phenomenon to the engineer.

Any attempt at chronology ceases at this point. My concern remains with the episodes remembered for their contribution to experience, not otherwise recorded, colouring subsequent views and impressions.

A particular inefficiency in the use of steel arch tunnel supports in rock arises from discontinuous blocking by means of timber stuffed between arch and rock, leading to bending stresses and lateral instability of the arch. The National Coal Board showed interest in solving this problem and provided full-scale test facilities to demonstrate benefits from innovation. A porous bag was placed between arch and rock (more than one bag where overbreak was excessive) and filled by pumping a weak sand/cement grout, excessive water issuing from the bag. The efficiency by comparison with normal timber blocking was raised by about 100%. The work was financed by an EEC grant and it is known that the method was adopted in European mines, but British safety regulations had been drafted such that the material for the bags, even though kept under water until filled, needed to satisfy stringent flammability requirements. Rock tunnelling in open jointed rock continues to use steel arches; the method of using bags for blocking was successfully adopted for the Cuilfail road tunnel at Lewes (Muir Wood 2000).

In the late 1970s, a programme for investigating sites for possible storage of nuclear waste encountered so much local opposition that the Government of the time confined consideration to sites already in public ownership, leading in due course to concentration on Sellafield. I had been struck by the absence of national co-ordination between the needs for deep investigations for other purposes, including geothermal energy, and the development of drilling and instrumentation methods. This led to the proposal for the Deep Earth Club which attracted interest from the Department of Energy, through the

Chief Scientist, Sir Hermann Bondi (and later Dr A. A. L. Challis) and the Geological Survey. In the 1980s, this notion of co-operation was scarcely the flavour of the political period and the attempt faltered. Perhaps the fact that we had different agenda did not help. One of my concerns was to make widely available all records of boreholes, as occurs in Northern Ireland (for boreholes deeper than 50 feet [15 m]) but not elsewhere in the UK. Hermann Bondi, on the other hand, was mainly interested in deep boreholes that might support the notion of continuous creation by evidence of helium atoms at great depth. I remain concerned at the commercial protection of borehole data, another fall-out of the 1980s. I await with apprehension an occasion of a serious ground failure on account of ignorance of information available from site investigation of an adjacent site.

I once again became involved in the disposal of nuclear waste in 1993. The Nuclear Industry Radioactive Waste Executive (NIREX) (UK Nirex Ltd), the body responsible for the disposal of radioactive wastes, approached the Royal Society to review the scientific basis for their work; I was invited to chair the working party (Royal Society 1994). We were impressed by much of the quality of the work we reviewed but criticised the unnecessary cloak of confidentiality under which much of the studies were conducted. Not only did this lead to external suspicions, often unjustified, but also to an inadequately systematic approach between scientists tackling individual problems and engineers who would need to synthesise the findings in planning actions. The greatest philosophical problem appeared to be with long-lived radionuclides which present an historical problem but need not be related to future nuclear power production. We were impressed by the Swedish programme and works, in the absence of such long-lived wastes. If this feature were more widely appreciated by the politicians, the historical problems need not be seen to cloud the decisions on future nuclear power, a nettle that needs to be grasped by a Government that does not prefer to postpone awkward policy decisions to successors – as time runs out.

I later had several experiences of appointment to consultants' boards, under different titles, including a major interceptor tunnel for São Paulo and for the Ocean Outfall Project at Sydney.

My work at São Paulo derived from an initial appointment to advise on the design of the tunnelling for the project, helping to weld together separate teams for soil mechanics, rock mechanics, and structural engineering and contract preparation. When work started, I was appointed chairman of a small team of consultants, which included

Professors Victor de Mello of São Paulo University and John Burland of Imperial College. We met at intervals of around six months to review experience and to advise on possible hazards ahead. From the outset, I had warned that the most difficult feature was presented by a back-filled creek on the line of the tunnel. Economic problems resulted in our disbanding. There was subsequently a major catastrophe when lives were lost as a Tunnel Boring Machine (TBM) became submerged in a soil inflow at this creek; I believe that this would have been averted if we had continued to function. It is difficult to justify a consultants' board if their presence is associated with no disaster; I have several experiences where I am confident that the advice of such a board, containing engineers with wide practical experience, would have avoided serious accidents that occurred in their absence.

Three outfall tunnels were to be constructed, advanced for several kilometres to sea, from Sydney's treatment works at North Head, Bondi and Malabar. The seven members of the consultants' board included expertise in geology, corrosion, the use of polymeric materials, hydraulic design of marine sewer outfalls, in offshore operations and in tunnelling. The Client authority, the Sydney Water Board, demonstrated a professional approach to the management of such an unfamiliar project which included several innovative features. Our visits were combined with presentations by the project team; this process in itself must have been of great value in developing an understanding overall of those working in the team. It merits record that, in the period following the one occasion of cancellation of our periodical meeting, incidents occurred both in the tunnel and the offshore works, not serious but which we believed we would have foreseen and helped to prevent. One of the subtler features of such a scheme with multiple discharge risers is to avoid a situation where one or more riser becomes filled with sea water whose relative density resists natural clearance. The hydraulic design has to strike a balance between efficiency over the life of the project with varying flows and the ability, at start-up, to blow out the sea-water intrusion.

6.7.2 Coasts, estuaries and landslides

The Leas Cliff at Folkestone, formed predominantly of the Folkestone Beds of the Weald clay, were moving, encouraged by lack of beach maintenance in wartime and undermining of the sea wall. An operation of urgent remedial work was interrupted by the Contractor's Insurers who averred that the work was hazardous. As Engineer, I disagreed

Learning from experience

and required work to resume. I agreed to meet the Insurer's assessor at site. Fortunately this was Sir Harold Harding, so he and I spent the day in discussing how to pursue the work with the greatest expedition, including, I recall, the use of a compressed air concrete displacer, a familiar device in tunnelling at that time, to blow concrete into the large cavities beneath the wall. I further recall that urgency did not allow tidal night-time piling to stop for a concert at the Leas Cliff Hall. The Resident Engineer reported that this caused some reasonable complaint, but that at least it fitted the tempo of the slow movement.

A footnote to Folkestone Warren, otherwise discussed in Section 6.5 above, concerns the several simple means for recording renewed movement of landslides, apart from surface surveying. Several of the timbered drainage headings were fitted in 1948 with movement indicators, whereby a length of signal wire was anchored at the in-by end of each, passing over pulleys to the entrance, where a suspended length of old rail kept it tensioned. Intermediately, brass arrows could be read against hardwood scales attached to the adit wall. This was effective in revealing the line of an active slip, even for small movements, the precise location of intersection helped by the records of samples taken previously at intervals along the adit to reveal details of the strata. When, in 1972, I was informed by the railway that movements had occurred further to the North of the part of the Warren that had already been stabilised, I enquired about the slip indicators. Their existence had been forgotten for about 20 years. On inspection, I demonstrated that they remained in operation and indeed provided precisely the information required. This surely demonstrates the virtue of simplicity – and of a system to retain a corporate memory.

At Lowestoft, the extreme eastern point of England, a massive vertical-faced sea wall had been underpinned, subsequent to its original construction, but was once again threatened with undermining by the sea. Judging from the shuddering response to breaking waves, collapse was not too far ahead. The objective was obviously to absorb wave energy while reducing risk of further material being washed from beneath the wall. The protection, therefore, needed to be designed as a series of filters, each size of material preventing scour of the next smaller size. The work had to be undertaken under water and exposed to unknown wave attack during construction. The solution was found in the use of gabions to contain the finer material, the cages obviously short-lived under abrasion and corrosion but lasting long enough for protective rock cover to be placed above and as toe protection. Surveys before and after illustrated a spectacular rise in the level of the sand beach in fine weather and after storms, the effect of reduced wave reflection extending several hundred metres offshore.

The notion of 'soft' coast protection obviously had more than one initiator. Once accepted as an obvious principle, examples were noticed from around the world. The traditional pattern of continuous masonry or concrete sea wall was often combined with a promenade and access to the beach for bathers and, at the zenith of the seaside resort, bathing-machines. Soft options, using to the greatest extent schemes for absorbing wave energy through the use of natural materials and natural beaches, needed to be based on an improved understanding of sand and shingle movements by the sea and the under-water sources of the materials. Were the updrift sources becoming depleted, obstructed by dredging, breakwater construction or by stabilisation works by others? The most satisfactory, and fairly sustainable, conditions for 'soft' solutions are found where geomorphology favours the formation of natural embayments. One such example in Thessalonica is described in Section 5.3. In some circumstances, little more may be required than the extension and protection of the natural promontories (where increased wave focusing may be expected in consequence) and infrequent beach replenishment as the natural material is abraded.

At Barton-on-Sea, there was no adequate natural embayment, so bastions were constructed at intervals as described in Chapter 5, to form bays between each. An initial problem here occurred as an urgent contract for stabilising the undercliff by drainage and regrading was about to be let. The Council discovered, at this moment, that instead of, as they had previously indicated, owning the land seaward of the cliff edge, their ownership only extended to the line of the cliff at an earlier date. Here was a dilemma! Should the work be redesigned to a less favourable profile, to be contained within this limit or should negotiations be started to permit the original design to be kept? The latter would take an indeterminate time. There was, in consequence, little option but to move the drainage works seawards, making stability of the lower under-cliff more difficult to maintain in perpetuity. This episode taught me that the inquisition from Engineer to Client should always include searching questions on land ownership to avoid such a problem.

Dungeness A nuclear power station was built on a shingle spit at the extreme south-east point of Kent. Shingle travelled from west to east under the influence of prevailing south-westerly winds and varying degrees of shelter. Hundreds of years ago, and the history of development can be readily reconstructed from mapping the shingle ridges from extreme storms, the shingle travelled around the promontory towards the north, the rate decreasing around the promontory which therefore advanced seawards. Later, a cusp developed and northward drift ceased, advance of the promontory increasing in consequence. The cooling water intake structure was to be constructed as a float-out piled tower. The question was, where? The requirement was for a minimum length of intake tunnel, the structure founded in optimal depth of water, ensuring submergence of the intake at all times, but sufficiently clear of the bed. For this purpose, sea-bed contours were prepared for the present day and as predicted for evolution of the foreland for 50 years hence (the intake might well outlive the initial station). The chosen site would be where the contours of present and future optimal depth intersected, checking intermediate conditions. As a related incident, during construction of the intake tunnel I was informed one day that the tunnel had encountered blackened timbers, evidently the result of a shipwreck. Soon after, silver spoons with Hanseatic league crests were discovered, readily dated. Comparison with the estimated seabed at the time at the point of intersection by the tunnel corresponded well, giving confidence to the estimates of off-shore profile related to the coastline.

An odd experience at around this time concerned a summons without indication of purpose to a meeting at the CEGB (Central Electricity Generating Board) headquarters at Bankside. I was confronted by a stern-faced inquisitorial panel who wished to know why the Board had been misled in advice on siting the Power Station at Dungeness. I knew well that this was not the case but, having had no warning of the subject of the meeting, I indicated that I would hear the accusations and respond outside the meeting. I had years of experience of the Board, including their practice of issuing minutes long after meetings had occurred, following internal circulation and 'improvement'. I therefore filed my own note and waited. In due course a doctored minute arrived, from which all charges had been pruned. I refused to accept this as a record and provided my own notes and refutation of the charges. I subsequently learned that, following our forecasts of the natural rate of recession of the foreshore along the frontage of the station, a building line had been prepared by the planning department. In transfer from one office to another, this line had been applied to the main reactor building and turbine hall, but not to ancillary buildings to seawards. Not being a party to this intention, I had assumed that countermeasures, of cost trivial in relation to the whole and probably in any event an economic solution, were intended. As a consequence a

Civil engineering in context

beach recharge scheme was set in hand, taking shingle from the promontory and depositing it to the west, to compensate for erosion. The whole operation was controlled by annual computation of volumetric shingle movement derived from aerial photography, later related more directly to calculation of volumetric gravel shift from annual storm records.

Subsequent studies at Dungeness have concerned the conflicting interests of gravel extraction and water supply from pumped wells. The tidal effects on salt-water intrusion were studied in relation to the distance between shore and ponds resulting from gravel extraction, also their water level. As the pond area increased, so did the natural rate of evaporation. There were also complications from the anisotropy of the deposits and the interests of wildlife, and the threats of weather and climate change.

When engaged in the protection of Brunel's Thames Tunnel in 1995, a question arose about the permeability of the Thames gravel above the tunnel on each bank. I suggested the sinking of two boreholes in a line transverse to the river bank so that water levels might be recorded over a tidal cycle for comparison with the tide. The hydraulic permeability could then be estimated by applying the equation:

$$\frac{\partial h}{\partial t} = \frac{T}{S}\frac{\partial^2 h}{\partial x^2} \qquad (6.6)$$

where

h: water level (m) at time t above mean level
S: storativity of gravel, i.e. porosity
T: transmissivity $\sim kH$, where H effective depth of gravel (m^2/s) and
k: hydraulic permeability (m/s)
x: horizontal distance (m)
t: time (s)

A solution for these boundary conditions, where $h \ll H$, is:

$$\frac{h}{h_0} = e^{-2\pi x/L}\cos(\omega t - 2\pi x/L) \qquad (6.7)$$

where

$$\frac{S\omega}{T} = \frac{8\pi^2}{L^2} \quad \text{and tidal period} = 2\pi/\omega \text{ (s)} \qquad (6.8)$$

h_0: value of h at $x = 0$, i.e. semi-amplitude of tide

Learning from experience

The only problem was that the record for only one borehole was provided, inadequate for the solution. Fortunately, Brunel had left with his records of construction, figures of water-table fluctuation in a shaft and an associated trial-pit; this at least provided a rough check for the order of magnitude.

A call one day in the 1970s from Cory Brothers, owners of docks, waterside buildings and other facilities, provided the information that their cellar at New Crane Wharf, a warehouse on the river's edge at Wapping, alongside Brunel's Thames Tunnel, had started to flood with each rising tide. The building was constructed by Thomas Cubitt in the early nineteenth century, with the footings sealed into the clay of the Woolwich and Reading Beds which underlie the gravel. The name New Crane Wharf was explained by an old map in the British Museum which revealed that the building was partially across the previous New Crane Dock, where masts were stepped into sea-going sailing vessels. Certainly, the water was rising and falling with each tide, with a phase delay. Water samples, however, revealed, particularly from their sulphide content, that the water was from a stagnant source and not directly from the Thames. It seemed highly probable that the Thames water was 'piping' through the clay seal that would have been provided at the mouth of the dock when it was backfilled beneath the building. A solution, other than pumping for perpetuity, seemed beyond reasonable economics. As a report was being prepared to this effect, a telephone call from the original source indicated that flooding had stopped as suddenly as it had started. Presumably, the 'pipe' had naturally sealed itself. We shall never know.

6.7.3 Silos and miscellaneous structures
It is peculiarly apt that the metaphor of 'silo mentality' is used as a pejorative description of engineers without interest beyond their narrow sub-discipline. This was precisely the feature that first impressed me from experience with working with silos. The three contributory disciplines of engineering: the structural engineer; the particulate materials engineer; and the materials handling engineer, had not only different and often opposed objectives but did not even use common symbolic language for the several parameters used in common. The design of the construction and operation of silos was a process calling out for a systems approach which, in the early days of my own association with silos, seemed absent.

Civil engineering in context

My most direct encounter with the problems arising from this lack of systems thinking occurred in 1969, following a spectacular failure of a new steel coal bunker of 2000 tonnes nominal capacity at Sharlston Colliery in the West Riding of Yorkshire. This was one of ten such bunkers constructed in 1967–8 for the 'merry-go-round' installations for providing coal by rail to power stations directly from the collieries. The bunker comprised an upper rectangular prismatic structure suspended from three portal frames, with a prismoidal under-slung hopper. Hydraulically operated gates at the base of the hopper controlled out-loading directly to railway wagons. The hopper had disintegrated and there was damage to the portal frames. The upper part of the bin remained essentially intact. The direct evidence was concealed beneath nearly 2000 tonnes of coal.

The first requirement was to establish the principal information, from examination of the site and records, relevant to the causes of the collapse. As the pile of coal was being recovered, an engineer posted to the site undertook an 'archaeological' record of the condition, relative position and orientation of each element of the structure as it was recovered, and removed for the evidence of failure to be thoroughly examined and photographed. In due course, this allowed a small-scale model to be made to illustrate the nature of the collapse. A curious feature of this investigation, never totally explained in the absence of systematic records of design decisions and inspections, concerned the many departures from good practice for a design-and-build project, including:

- The length of each of the 42 supporting bored piles (apart from the first) had been reduced in length from 18 ft (5.5 m) to 13 ft (4 m) below base of footing at an unrecorded site meeting. There was minor but insignificant differential settlement.
- The hopper was lined in glass tiles to reduce wall friction. This has the effect, illustrated by Fig. 6.4, of increasing the bursting forces against the hopper. The specification was nevertheless based on Rankine active state loading.
- While calculations were available for each of the basic beam and plate elements of the structure, none could be found for the corner junctions, where secondary stresses would be appreciable.

Although each of the above features might add to overstress of part of the structure, collapse would not have occurred but for a more remarkable discovery. The sides of the hopper were designed to be joined by $\frac{5}{8}$-in (16 mm) steel tie-plates between outlets, through the

Learning from experience

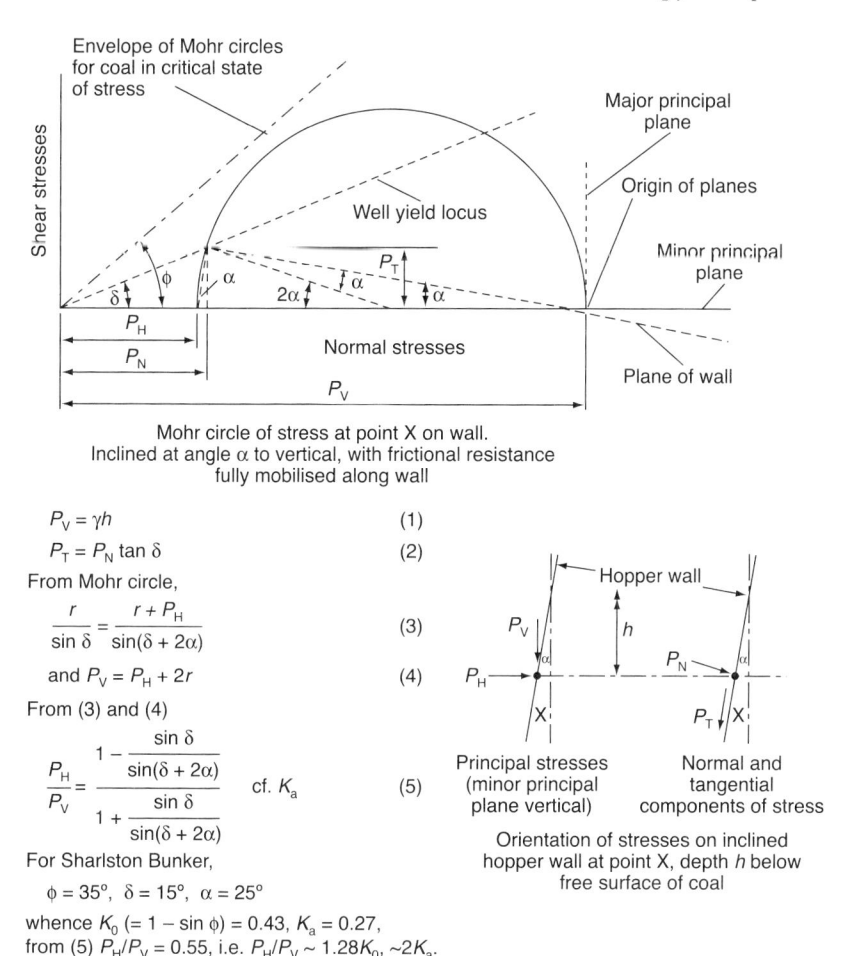

Fig. 6.4 Bursting pressures for glass-tile lined silo hoppers

Mohr circle of stress at point X on wall. Inclined at angle α to vertical, with frictional resistance fully mobilised along wall

$$P_V = \gamma h \quad (1)$$
$$P_T = P_N \tan \delta \quad (2)$$

From Mohr circle,

$$\frac{r}{\sin \delta} = \frac{r + P_H}{\sin(\delta + 2\alpha)} \quad (3)$$

and $P_V = P_H + 2r \quad (4)$

From (3) and (4)

$$\frac{P_H}{P_V} = \frac{1 - \dfrac{\sin \delta}{\sin(\delta + 2\alpha)}}{1 + \dfrac{\sin \delta}{\sin(\delta + 2\alpha)}} \quad \text{cf. } K_a \quad (5)$$

For Sharlston Bunker,

$\phi = 35°$, $\delta = 15°$, $\alpha = 25°$

whence $K_0 (= 1 - \sin \phi) = 0.43$, $K_a = 0.27$,
from (5) $P_H/P_V = 0.55$, i.e. $P_H/P_V \sim 1.28 K_0$, $\sim 2 K_a$.

Orientation of stresses on inclined hopper wall at point X, depth h below free surface of coal

---

agency of a continuous ring beam. A full-strength weld was required at each such junction. While there was evidence of a trial shop assembly to establish good fit, errors in construction had resulted in a poor fit at site between the tie-plates and the ring beam. One end of each of several tie-plates had been burnt off in consequence and the weld reduced, in certain instances to the introduction of steel rods with tack welds only. Above each tie-plate, inclined steel fairings were attached to each side of the hopper. The tension through these non-structural features had evidently held the structure together until the collapse. Damage to the portal stanchions was found to be consistent with the kinetic energy of the hopper sides as, at failure, they rotated about a top 'hinge' and struck the stanchions, prior to falling free. While the indications of poor quality control in design, construction and inspection were obviously of greatest concern to those involved, it is the question of side-wall friction which merits a little more discussion. The materials handling engineer requires lowest friction, the structural engineer is seeking most economic design. The feature illustrated by Fig. 6.4 of pressure in excess of the 'at rest' stress would have been evident to an engineer familiar with the behaviour of particulate materials. The angle δ relates to friction between the coal and the glass-lined wall of the hopper, ϕ to the angle of internal friction of the coal, K_a active Rankine pressure factor, K_0 at rest pressure factor. Curiously, the booklet in the UK available at that time on the structural design of bunkers and silos (BCSA 1968), issued by the British Constructional Steelwork Association (BCSA), makes no mention of this feature. Lightfoot and Michael (1967) indicate that the CEGB used Rankine active state design at that time. More curiously yet, the discussion by Mr R. H. Squire on a paper presented at the ICE (Bridges 1951) on the design and construction of silos and bunkers, raised this precise issue; the author, in reply, contradicted this view and the record was not further challenged at the time. Hence, several published accounts available to the structural engineer at the time of the Sharlston collapse were misleading. This is clearly the result of a failure in communication between the particulate engineer and the structural engineer.

Subsequently, an opportunity to chair a small working group set up by the British Materials Handling Association, provided a simple 'nesting' series of design guides, each aimed for an appropriate level. For example, for the simplest small symmetrical silo, structural design is often dominated by provisions for corrosion and durability, so stress levels may be small. At another extreme, the large silo with asymmetrical outlet needs a thorough analysis of the most severe dynamic loading. The silo may later be used for a material different from that for which it was originally designed; this possibility should not be overlooked. Much work on silos has progressed since this time, with a comprehensive study by Rotter (Rotter 2001) providing a handbook for circular steel silos, which links the several contributory engineering disciplines, taking account of the behaviour of the contents, including the pulsating stresses as out-loading occurs. There is a statistical aspect to these stresses, so that the maximum load is related to the critical area under load from the structural viewpoint.

Learning from experience

Horace Morgan had been associated with Barnes Wallis on the 'bouncing bomb' used in the Second World War to attack the Mohne dam. The bounce depends on the hydrodynamic forces created by a 'reverse' rotation. This is another not obviously intuitive result. Many years later, Horace Morgan had been approached by Barnes Wallis with the notion that geodetic construction would be the economic way to design an above-surface railway. I had previously encountered geodetic construction in my vacation work at Vickers, building Wellington bombers. The structural members follow diagonal geodesic curves, i.e the surface in the shortest line between points of intersection, which introduces a measure of redundancy in the structure. This was found to be beneficial in reducing effects of war damage but the benefits for a peace-time bridge structure seemed less obvious. However, a short section of a possible rail route to Heathrow Airport was studied, a geodetic structure with longerons compared with a more conventional approach, drawings and rough estimates prepared; the proposal was found to be unattractive. Dr Wallis was unimpressed and insisted that any one of hundreds of his 'stress-men' would be able to demonstrate the merit of the approach – but that none had time to do so.

In 1972, a joint appointment by British Railways, London Underground and the BAA provided the opportunity to plan a joint rail access to Heathrow Airport, by rail from Feltham on the line from London to Staines, and by underground by extension of the Piccadilly Line from Hounslow. These were seen as complementary links, the former predominantly for air travellers, the latter for those who worked at the airport. Planning was difficult, since the primary object of progress meetings, to record progress and determine the next steps in the process, was thwarted by the fact that railway departments did not talk to each other between meetings. Stephenson's precept (Stephenson 1856 p. 144) was yet inadequately heeded: 'no railway can be efficiently... conducted, without thorough unity between the heads of all the great departments.' Nevertheless, progress was made, if slowly, and the outline of a scheme with a joint station took shape. A system was devised for electronic checking of baggage at the London terminal, Victoria or Charing Cross, stretching the available technology of the time. Estimates of cost were prepared, including the track, control and operating costs provided by the railways. These latter seemed a little light for London Underground, and to include other desirable resignalling work for the Southern railway, so were scarcely comparable. They were deemed to serve for a first overall assessment.

Civil engineering in context

Meetings with the Ministry concerned prepared a basis for a parliamentary Bill. Unfortunately, the timing was wrong, in Edward Heath's 'abrasive year' as Prime Minister. It was declared that the cheapest 'option' – but these were complementary – should be preferred. As work on the Piccadilly Line extension subsequently progressed, so was space left surreptitiously for the possible reversal of policy. Development to the south of the airport soon made this route impracticable, with obstructions to the lines of bridges and tunnels to cross roads and rivers.

In the 1980s, an approach came from Peter Coni QC, Steward, Henley Royal Regatta, to develop *pro bono* a scheme for a pedestrian crossing of the River Thames at Henley for the Regatta. A subway beneath the river was obviously out of the question, with the high costs of maintenance throughout the year. What seemed much more attractive would be a bridge erected and dismantled each year as an exercise by the Royal Engineers. Much support was obtained for the proposal for a demountable suspension bridge, with ICI prepared to supply the parafil cable, Halcrow to design the bridge and others to provide the structure, all for trivial cost. The piers would be sited on the line of the markers defining the racecourse. What remained was to obtain the agreement of the Chief Royal Engineer to the annual exercise. This was immediately attractive as a valuable experience and one that would receive good publicity. Unfortunately, Margaret Thatcher as Prime Minister had recently decreed that the Army were to recover their costs of any such operation, a lack of imagination that could not be avoided. So perished this chance to explore new ideas and materials in a way to allow improvements year by year in the light of experience, to be applied to suspended structures on a larger scale. The recent example of erection of half-scale ribs for a model of Ironbridge by the Royal Engineers implies that this short-sighted decree has now been lifted.

7
The engineer for the twenty-first century

7.1 The historical legacy

A group of actors has reduced Shakespeare's plays into a single evening's performance. In comparably cavalier style, Chapter 1 cantered through a review of some of the engineering achievements of the Industrial Revolution. There were then perceived needs on the one hand, unfolding understanding of the possible applications of science on the other. The engineer was largely concerned with meeting demands, sometimes ahead of an adequate understanding of the science behind the technology, sometimes deliberately building upon a science, perceiving the latent capabilities. In all respects, however, the objective was to satisfy a growing market, the market influenced by perceptions of benefits by improvement of the standard of living, the industrial needs of denser centres of population, of transport, people and materials, coupled with the opportunity for personal profit. There are, of course, many aspects of the relationship between the engineer and society. For the sake of the present context, it is necessary only to emphasise that the engineer was perceived as possessing the skills necessary to apply technology – generally at that time perceived on balance as beneficial – to improvements in society. There were some who would have preferred to live in an increasingly distant Arcadian society, or at least said that they would, as the old ways were displaced. These were a small, if articulate, minority with a natural sensitivity in resistance to change.

The role of the engineer was accepted as adventurous, engaging in technological forays without precedent and without adequate experimental or theoretical evidence for success. Instead, however, of urging caution and restraint until technology (a word not current) provided an adequate basis for confidence, pressures continued to be applied to engage in further extrapolation of experience.

Two features deserve notice. First, that when things went wrong, as happened fairly regularly, sanctions on the responsible engineer were not pressed hard home. A remarkable example is provided by the fatal accident caused by the collapse of the Dee Bridge for which Robert Stephenson was responsible. This was recognisably, even at that time, a poor design of a cast iron girder bridge, reinforced with wrought iron rods inadequately anchored. Curiously, Stephenson was supported by other eminent engineers (Rennie, Vignoles and Locke) which doubtless overawed the Royal Engineers officers who had investigated the accident (Walker and Simmons 1847). There was an open verdict; however, similar bridges were redesigned.

After the hectic pace of the mid-nineteenth century, technology settled down in the early years of the twentieth century, depending upon construction and manufacturing techniques developing incrementally but without major repositioning of the players. Thus developed the several engineering disciplines related to their respective industries. Traditional techniques appeared adequate to meet demand, with an expanding empire providing new markets. While manufacture concentrated increasingly on the processes, construction applied its major effort to the product. The engineer was generally distant from the market.

Britain appeared to be a mature market with demand for new work inadequate to promote innovation. British engineers were increasingly in demand overseas where the state of development expected fairly primitive, robust methods of construction and hence overall simplicity. Engineers undertaking works for the armed services, for example through Navy Works, were well placed to understand the specific requirement of their client and learn from experience in successive works of comparable nature. We see a similar development of the US Corps of Engineers and Bureau of Reclamation, until resistance to external influence bred excessive self-confidence.

The engineer tended towards complacency in an unchanging set of procedures, largely dependent on precedent, the complexity of the engineering controlled by the limited means of computation. The engineering surveyor only needs to be reminded of the volumes of 8-figure log tables! At a personal level, I have recalled in Chapter 6 experience in working through the shops of a shipyard on being commissioned into the Royal Navy in 1942. My impression was that the technology had not changed since I lived in Chatham Dockyard around 1930. On further reflection, it seemed that little had changed since around 1900. The ships had changed but not the way they were built and fitted out.

The engineer for the twenty-first century

Internationally, there were stirrings as new demands set new problems. For example, the Permanent International Association of Navigation Congresses (PIANC) was founded in 1905, the Permanent International Association of Road Congresses (PIARC) at much the same time. Both organisations organised periodical congresses at which a set (for PIARC, for many years the same set!) of technical questions was posed to the members, in an attempt to share good practice in the design and construction of roads. Only in the last 20 years, has PIARC developed a major interest in the social and environmental aspects of road transport, influenced initially by its Technical Committee on Tunnelling where these aspects were recognised as vital factors in the decision to go underground.

The research needs of a largely empirical industry of the early twentieth century were addressed centrally through Government funding, principally from the viewpoint of achieving satisfactory products of national interest. As road traffic began to develop (Charlesworth 1987) a Road Board was set up in 1910, financed by motor tax revenue through the Road Improvement Fund, making use of part-time technical consultancy. In 1919, the Board was replaced by a Ministry of Transport (MoT). In 1925 the Road Improvement Act was concerned also with the improvement of road construction. Work was already commissioned through the National Physical Laboratory (NPL) on questions of skid resistance. In 1930 an Experimental Station was set up on Colnbrook bypass near Heathrow under MoT engineers, in 1933 transferred, following a report of a House of Commons Select Committee, to the Department of Scientific and Industrial Research (DSIR) and renamed Road Research Laboratory (RRL). Responsibility was divided between the MoT, carrying out experiments on public roads, and DSIR, concerned with research. DSIR was closed as a result of the Rothschild Report (1971), 'A framework for Government Research and Development' based on the customer/contractor principle for research management. RRL moved to Crowthorne. The name was changed to Transport and Road Research Laboratory (TRRL) and later to TRL, and subsequently privatised to TRL Ltd.

During the First World War (1914–18), a building materials research committee was set up, transformed to the Building Research Board, with the object of improving knowledge of new materials, physical processes, the behaviour of buildings, conveying information to the building industry and to the public. It was established first at East Acton and Farnborough, moving to Garston in 1925, later merging with the Fire Research Station at Farnborough and the Forest Products

Civil engineering in context

Research Laboratory at Princes Risborough. Reginald Stradling, first Director of the Building Research Station (BRS), was also appointed as Director of the Road Research Laboratory in 1930.

For many years, consulting engineers in Britain had campaigned towards an adequate facility to undertake research for free-surface hydraulic problems of inland and coastal waters. In 1947, Sir Claude Inglis was appointed director of research for the UK Hydraulics Research Board. The first stage of construction of the Hydraulics Research Station (HRS) at Howbery Park, Wallingford, was completed by 1955. Initially, much of the work was undertaken in large-scale open-air models. Sir Claude brought with him the empirical regime theory of Indian waterways, to which he had contributed over many years, and his experience with similar models from the central research station at Poona, which he had founded.

In the 1950s there was a perceived need for a research and development organisation based on the construction industry, concerned more with the processes of design and construction than the products. Construction products constituted the principal interest of BRS, RRL and, later, HRS. The Civil Engineering Research Council was formed in 1960 from the earlier ICE Research Committee, following discussions with DSIR and industry, with increased support from industry, the ICE and the Federation of Civil Engineering Contractors. Sir Herbert Manzoni, then the President of the ICE, was appointed as first Chairman of Council. In 1962, the first Director was appointed. In 1964, the Civil Engineering Research Association (CERA) was formed with a Research Advisory Committee. The name was changed in 1967 to CIRIA, with increasing funds from members, predominantly from industry. There was a special initiative for piling in 1974, and with concrete in the oceans (1975), an element of the specialist offshore community, the Underwater Engineering Group. Many subsequent initiatives have been established under the several broadening activities of CIRIA, many of these in areas of environmental engineering and sustainability.

With an increasing graduate entry, the profession became increasingly technocratic, with greater interest in the application of science than in the promotion of projects; at the same time, engineers began to use more technical jargon, speaking languages not readily understood by the layman. Thus arose a largely isolated profession, not yet adequately concerned with the external significance of projects, hence lacking any special competence to discuss social and environmental aspects. As the benefits of the infrastructure and commodities provided by engineering became increasingly accepted as a natural way of life, so did the attraction increase, particularly by social scientists, of drawing attention to the negative aspects of industrial society. In this way the 'green' movement arose in the 1960s, stirred also by political events of the time, which identified selectively the down side of industrial development with its inadequate concern for consequences. At least in part, this 'green' activity represented an overdue need for a more holistic (not a current word of the time) approach by engineers, hitherto deficient in their education, training and, too often, terms of reference for a specific project. Unfortunately, among too many teachers and manipulators of the media to influence public opinion, the negative picture, more readily grasped by those without technical knowledge, received excessive publicity, without regard for balance. The relationship between services expected by the public, their cost and the social and environmental consequences was too little emphasised. The essential balancing exercise in policy decisions was ignored. In consequence there arose an increasing gulf between the 'greens' and the embattled engineers who found that they were subjected to an emotionally, if quite illogical, contrived attitude for 'anti-technology' (Florman 1976).

The positive achievements of the engineer were submerged in the public view beneath any cause, or suspected cause, of environmental damage. Engineering contributions to the needs and diversions of people were taken for granted. A technological age, with increasingly democratic power over machines, causes perplexity for those, too many and probably increasing, denied a grounding in science. Thereby arises a new variant of C. P. Snow's social division, not now confined to the academic divide between the arts and the sciences but, more broadly, between those who attempt to understand science and technology and those who do not. The latter turn to the therapists and the creationists, the witch doctors and false magic of an uncomprehending society. The media, who largely do not understand science, then find much satisfaction in publicising trivia of the odd-ball research interests of the scientific or pseudo-scientific world as if these were representative of the central work of engineers and scientists. In a democratic industrial society, or one that continues to live on the fruits of industry, it is imperative that the basic understanding of science is seen as an essential part of education. The alternative would entail lowering standards in all those qualities of life that people consider their right or expectation.

The engineer, cocooned in his narrow definition of his preoccupation as a single discipline technology in the planning and execution of

Civil engineering in context

projects and processes, but not in the directing policy, became increasingly isolated and defensive. The very choice of a sharply focused career tended to attract the technocrat without interest in, or ability for, communication outside his discipline. The architect, by comparison, although sharing much of the engineer's isolation from policy, maintained a far greater visible facility for creative ability and hence attracted better communicators who, away from aesthetic theorising, spoke a simpler language than the engineer.

Fundamentally, the ethical essence of engineering (see Section 9.2) appeared lacking in a profession which had no formal training in risk and the perceptions of features beyond the efficiency of the individual project or artefact, measured by internal financial cost/benefit. Too readily, the engineer, particularly in the role of management consultant, became identified with the caricature of super-national business trampling the interests and desires of the less powerful.

7.2 Repositioning of the civil engineer

I am able to live with the terms of the definition of the civil engineer as one 'who directs the great sources of power in nature for the use and convenience of man' attributed to Thomas Tredgold (1788–1829) in its broadest definition. I interpret this as representing man of an age of enlightenment capable of recognising the duties and responsibilities coupled with reconciling wider interests with his own 'use and convenience'. I have sympathy with those who see little sign of such insight from a society whose external features of greed appear to overweigh sensitivity and altruism. The civil engineer needs to balance abilities to provide and maintain the infrastructure on which the modern world depends, while contributing to the perception of conflicting consequences and their mitigation. Increasingly, it is necessary that decisions on policy take account of features that require the civil engineer's capabilities for their perception and definition.

The inevitable conclusion is that the civil engineer's profession needs to embed the specific and specialist aspects, the traditional structures, geotechnics and hydraulics within a culture which extends to the understanding of risk, and awareness of the possible side-effects of civil engineering works, and their mitigation. The underlying essence of the civil engineer's training, as for all varieties of professional engineer, is that of working in systems (see Chapter 5). Once the notion of a system is understood as the underlying capability of engineering, the engineer as communicator is demonstrated as an essential element towards operating the system, in which the concept of risk plays a dominant part. Risk may be defined as broadly as we wish for an individual project, dependent on its nature and scale, but social and environmental impact are clearly all part of the risk.

An individual engineer is not conceived as a polymath but as one who has developed adequate capability to communicate with those near to his field of responsibility, able to complement his own understanding, as illustrated by Fig. 5.3. In the hierarchical scheme of a career, the area of communication will expand, accompanied by greater dependence on others with more detailed and up-to-date skills in the most specialist zones of the engineer's work. There will be some who concentrate on the development of greater understanding in specific areas, by research and applications. Many others will extend their areas of understanding, an ability depending more on experience and exposure than upon formal education. The most competent engineers will be able to fit their specialist knowledge directly into its wider context.

Engineers owe a responsibility to another part of the system, that of the attraction of talent, its education, training and application. These are all highly interrelated activities. Evidently, the greater the civil engineer's ability to influence policy towards beneficial consequences and efficiency in the procurement of resources towards this end, the greater the expectation of positive influence towards recruiting those with the highest abilities. There is great personal satisfaction to the engineer engaged as one of an identifiable team in a well-conceived project. The degree of all round success of the project will, incidentally, be reflected in the degree of recognition and reward for the engineer's contribution – another part of the system.

The proposed system of broadening and unification of the engineering profession, through the notion of colleges, discussed in Section 3.2, provides for the option of qualifying as civil engineers those who engage in applied sciences allied to civil engineering but not traditionally part of the civil engineer's curriculum. Obvious contenders would include, for example, those concerned with the biological treatment of wastes, water storage in geological formations, disposal of nuclear wastes, asset management in general, all aspects which entail the application of science towards the ends encompassed in the civil engineer's role.

During 1981–4 I had the privilege of serving on the Board of ITDG (Intermediate Technology Development Group), founded by E. F. Schumacher. The most interesting part of the work of the Group, to me, lay in the identification of appropriate means of solving problems in one developing country and passing the technology, doubtless with modification, to others. I had much admiration for those who devoted their lives, with modest recompense, to this work, of whom many were, and are, engineers. Subsequently, I was invited by the Institution of Mechanical Engineers to open the discussion following a lecture by Professor Meredith Thring (author of *The Engineer's Conscience* 1980) on the general subject of appropriate technology and green issues. I opened my contribution by expressing my sense of honour in this invitation for a prestigious lecture, especially as I was not a Member of this Institution, and most especially since my application to join had been rejected. This caused considerable amusement to the meeting, which included the Prince of Wales who also contributed. (I was later told that, at the time of my application in the early 1950s, job title was the key, not experience; this made sense since I knew I had considerably more mechanical engineering background than my principal supporters!). My constructive suggestion was that engineers too often found that they were unable to develop appropriate solutions, since the economists had been there first and already chosen centralised, inappropriate technology procedures. I recommended the inauguration of the Appropriate Development Forum, which would welcome economists and other social scientists, and others, to join the engineers, with an interest in helping developing countries. This initiative was taken up by the Institutions of Civil and Mechanical Engineers.

In the 1970s, it became apparent to those working overseas that international disasters indicated that medical assistance needed to be complemented by work by engineers to restore services and provide tolerable living conditions. An initiative by Peter Guthrie and others led to the formation of the Register of Engineers for Disaster Relief (RedR), supported with office services by the ICE. A register is maintained of those willing and qualified to be called upon to work in an appropriate manner in a disaster area. When the occasion arises, calls are made to find those suitably qualified who are available within the time frame.

7.3 Satisfaction as a career

The civil engineer is fortunate, at least in principle, in expecting to gain the main element of job satisfaction through the knowledge of contributing to a well-conceived goal achieved through a high quality of elements and their synthesis, essentially as a well-managed system. It

The engineer for the twenty-first century

is then reasonable that any engineer, allotted a particular task to perform, should understand how this fits into a wider objective. The hierarchical system should be accessible, possibly in a simplified form for a developing project, through a display, with each sub-system (and sub^n-system), presented in greater detail for those working within it. Not only does such a method of briefing mobilise the interests of the engineer but it enables him to perceive more readily where any particular form of solution to a problem he has been set may interact with another aspect of the whole.

The experience of working in a particular group will widen the understanding of its work and of its interactions within the hierarchy and the cross-linkages. Not only the individual, but also the quality of work, benefits from changes in activity, encouraging a firmer grasp of the adjacent elements of the system. Increasing experience and seniority, of course, entails greater breadth or depth of understanding, preferably both.

A commonly asserted glib notion asserts that a constant change of activity and employer represent a universal feature of success at the present day. This is a banality which appears to gain acceptance from its constant iteration. In the commercial world, short-term financial benefit comes from maximising immediate profit to the individual in a world without long-term perspective. Hence, the individual has the urge to suck the nectar of reward from each flower, heading always whither, in the short term – to vary the botanical metaphor – the grass appears greener. His experience is valued more on degrees of insider knowledge than capability, so that even the threat of seeking a new employer may achieve the intended consequence. Even in the commercial world, this life style seems economically unsustainable and has only survived to the present by connivance between insider interests and the Inland Revenue.

The longer focus of the professional coupled with different sources of job (and life) satisfaction demand a different set of values and hence solutions. The greater degree of symbiosis between the interests of the organisation and its members confers a benefit upon stability of occupation where mutual satisfaction does not lead to stagnation of enterprise. Two-way loyalty, a foreign notion in the free market, is at the centre of the relationship.

It is to the benefit of the organisation and the individual for periodical exposure to debate on issues of particular or more general interest, internal or external to the organisation. This was a principal role of the early engineering institutions. As one of several examples I have already recalled, experience as a member of the consultants board for the Ocean Outfalls Project of the Sydney Water Board (Section 6.7.1) between 1985 and 1991. A contagion from the commercial and legal world is the notion of virtue demonstrated by working excessive hours and applying excessive proportion of effort to direct revenue earning as against attention to best practices, once again prompted by short-term issues. In the engineering world, the driving factor for this distortion is the excessive competition on price, which promotes fragmentation of responsibilities, inadequate attention to risk, absence of effort towards the optimal solution – and many other undesirable consequences. Time for reflection, and time for other interests, is as important for the life of the engineer as it is to accumulating the capital of corporate know-how, the essence of the surviving organisation.

The first engineering appointment to a project may well determine the extent to which quality will bear upon subsequent appointment, by whatever means. If the first appointment is by price and in consequence placed on the defensive against other participants, the remainder will follow suit and the whole will suffer. Chapter 4 has emphasised the benefit from there being no element of competitive firm cost in the first appointment. The first appointed engineer should reasonably provide an estimate and a basis for computation of fees and expenses, within a system for periodical reporting and sanction. The argument for not declaring a firm cost is overwhelming in protecting the client against risk. Risk will be minimised by an overall policy of risk management, not by deflecting liability on to others, an inevitable consequence of the first player needing to minimise his own costs. In a perfect world, only those prepared to work in a totally responsible manner would be charged with such activities. In a world conditioned by the excessively commercial attitudes recently embraced, it would be over-sanguine to expect that those, conditioned by this climate, would be suddenly converted to the virtues of a professional outlook. Examples will spread the belief more effectively than precept. It deserves note that an initial appointment by competition on price will also impair the standard of care for the social, environmental and sustainability features.

Chapters 2 and 5 each refer to the notion of the intelligent market, a broader version of HM Treasury's definition of the 'enlightened purchaser'. The essence of the intelligent market is that it provides a result in rewards (the game theory win-win situation) for all concerned greater than could be achieved by the strict application of the free market, where each item would be sought (but rarely obtained) at least cost, dissociated from the others. There are several aspects to the intelligent market. The initiator of a proposal for private profit would thereby consider long-term and possible external benefits of potential interest to the public sector, with the object of sharing costs and benefits. Variations would be considered to match these interests. As such a scheme developed, so would the planning be undertaken with a deliberate objective of continuing the process of optimisation overall. Questions of the overall system of procurement, and of processes for dealing with risk, would be set up and carried through the resulting project. One feature of the intelligent market is that policy is directed throughout by those with an understanding of the practical consequences of each decision, in preference to reliance on a simple set of markets to deliver each element at least cost. The difference between the intelligent market and the Latham and Egan Reports is that these reports are relying on attempts to improve the performance of the construction industry (the market) solely (Latham) and predominantly (Egan) by commercial mechanisms. In this way, they are essentially limited in their disregard for the professional element and the benefits derived from the cohesion of a system intelligently operated. The intelligent market represents a conscious attitude by all the parties concerned to work towards their mutual benefit; it does not imply any specific set of relationships between them. Within the project, the intelligent market has the virtues of partnering. A number of success stories recounted in this book illustrate the benefits to be attached to this approach.

An important factor in overall satisfaction to the individual and to the profession concerns the relationships between the parties engaged in a project. The objectives of a Contract or Agreement are to define responsibilities and terms for reward. When, on the contrary, the Contract becomes a defence treaty for the Employer, war has been declared and the engineering suffers with the Project. Chapter 4 describes examples of good practice. These should become universal. All will benefit, whose objective is good engineering and value for money.

7.4 Engineer's code of conduct

Fundamental to the work of an engineer, as of any professional, is the ethical code under which he works. Wider aspects of ethics are discussed in Chapter 9. The engineer will only be able to work fully within an ethical code under a regime corresponding to the intelligent market described above. The engineer exercises his professional duties by applying creative energies to the development and application of concepts based upon technical knowledge and know-how gained through experience. The technologies that he presently commands are based upon a far wider body of science than his counterpart 200, 100 or even 50 years ago; his essential function remains that of judgement. His professional judgement may often need to be applied where there is less than complete certainty as to the outcome, a consequence that the engineer should always discuss with his client, i.e. whoever rewards the engineer for his work. In this respect, he should reflect on the interpretation to put to Rule 1 of the 'Rules of Professional Conduct' (ICE 2000a): 'A member shall discharge professional responsibilities with integrity and shall not undertake work in areas in which the member is not competent to practise.' From time to time, an engineer finds himself working in an unfamiliar field because of the novelty of his approach; he then has to decide whether he needs on this account to work from first principles or whether, on the contrary, there is a body of knowledge to be drawn upon from elsewhere. As so often in engineering, he needs to know what he does not know.

The engineer engaged in preparing codes of practice should always consider the different levels at which these may be applied. At a simple, small scale, more severe limitations may be applied than for greater complexity or size where a greater amount of skilled input will be justified to lead to a modified but adequate set of conditions. The different levels should 'package' one within the other to avoid confusion from 'picking and mixing'. Their scope and level of appropriate understanding should always be made clear.

In logic, all those who work as technologists, applying science to a practical end, should be expected to share a professional duty with the engineer. It may be argued that some applications of science entail little intervention of judgement and hence that moral duties extend no further than probity. The proposals for 'colleges' of Chapter 3 would provide an opportunity for qualification as engineer of many applied scientists who would then support the professional declaration of the institution to which they were attached.

Rule 3 of 'Professional Conduct' reads: 'A member shall have full regard for the public interest, particularly in relation to the environment and to matters of health and safety.' Rules 1 and 3 basically establish the duties to the Client and to society. For a detailed account of engineering ethics, the reader should turn to Armstrong *et al.* (1999). Rule 3 also introduces risk and the general principles of risk management as an essential part of his culture, to enable him to foresee and explore, often with others with different expertise, the potential consequences of a proposed development, now formalised as Environmental Impact Assessment. The terms of reference to an engineer's activities may exclude environmental issues, as being undertaken by another agent. It remains, in such circumstances, the engineer's duty to ensure that aspects foreseen as having a significant effect are not being overlooked.

There is a strong tradition in civil engineering, particularly in the public sector, to consider the wider public interest. Reorganisation of public services has transferred engineering posts to those without commitment to a formal set of professional rules. In consequence, it may be more difficult for the engineer working at mid-level to ensure compliance with Rule 3, apart from issues conceived by his seniors as affecting the 'image' of the organisation. In the private sector, the form of ownership of the firm only becomes a problem if shareholder interests constrain the professional from acting wholly in the interest of the Client, and of the public in relation to social and environmental issues.

By a currently proposed simplification of the rules of professional conduct, the remaining requirements are for an engineer to respect the environment and 'the sustainable management of natural resources', a wide responsibility until boundaries are defined. The engineer would also be required to notify the institution if convicted of a criminal offence, declared bankrupt or disqualified as a company director.

From time to time there are calls for registration of engineers. Several proponents of the Finniston Report (Finniston 1978) were keen on statutory registration, effectively removing the self-regulating function of the engineering profession. This attitude was largely by political motivation, manifesting the politician's reluctance to share his privileges with professions, an attitude discussed further in Chapter 9. Others would like to see certain tasks confined to engineers 'registered' by the institutions. In so doing, the Reservoirs Act 1975 is quoted, which requires periodical inspections of each dam by an empanelled engineer. For this relatively definable area, the panels relate the significance of the reservoir to the experience of the engineer as inspector. Any attempt to provide a scheme with more general application of any significance – and protection of the public must be the primary aim – rapidly becomes hopelessly complicated. How many categories of registration would be needed between those who design forest tracks and those who design motorways, not forgetting those who know the best methods of constructing each? In the US, where registration, by each State, was introduced largely as a result of the numbers of immigrant engineers from so many different cultures of education and training, the registered engineer has the function of approving with his signature work undertaken by others. The registered engineer, therefore, has to have adequate confidence in the competence of the unregistered engineer for whom he accepts responsibility, effectively undertaking a secondary degree of control. Regard by the engineer for his institution will be enhanced by convergence between support required and services offered, not by a heavy-handed and, incidentally, politically unacceptable, registration of engineers.

With our own tradition of qualification by examination and interview, an alternative means is available, providing these functions are undertaken competently and thoroughly, offering a genuine protection to the public. Acceptance to membership following qualification is associated with undertaking to comply with the Professional Rules. At all times, the engineer should be reminded of his undertaking not to engage in work outside his competence. I have stressed above that the engineer must know what he does not know, to equip himself for recognising his boundaries for competence. This feature should be rigorously policed by the institution; the assurance of compliance should be a selling point of a member of the institution, a 'kite mark' of quality assurance to the public. In this way, esteem for the engineer will accompany a growing respect for his professional institution.

7.5 Implications for education and training

Implicit in the definition of the activities of the professional engineer (Section 3.5) is that he will be expected to work beyond the limits of prior experience. This must imply the ability for learning in a working environment and to take decisions on the basis of background knowledge and judgement. The primary object of his education must be to enhance these capabilities. Charles Inglis (Inglis 1941) quotes approvingly from Dr Arnold thus: 'it is not knowledge but the means of gaining knowledge, which I have to teach'.

Chapter 5 has established the need for the engineer to understand the principles behind the solution of problems by computer methods. There has been much discussion on the engineer's need for mathematics, the argument obscured by failure of the proponents for reduced mathematics to define what these reductions should be in scope and depth. A sub-argument concerning whether the engineer needs A-level mathematics is a question for those universities prepared to organise classes in remedial mathematics, within a degree course, without loss of quality, for those without adequate grounding. The move towards the International Baccalaureate, with a core containing mathematics and English, should help to remove this specific issue. One of the most articulate but protean debaters of the subject asserts that the need for mathematics deflects some of the most creative minds from engineering. The challenge to establish the existence of a negative correlation between mathematical aptitude and engineering creativity has not been answered. It seems far more probable that these creative minds have been repelled by mathematics taught as a chore by those without enthusiasm, aptitude or knowledge of application. The academic world has criticised the Joint Board of Moderators (for the ICE and Institute of Structural Engineers (IStrucE)) for allowing approved courses to accept candidates without A-level mathematics. The Board in reply has suggested that this relaxation should only apply to IEng courses, implying that the purpose is to encourage a wider entry and not apparently related to the question of relating innumeracy to creativity, which might appear to relate more directly to the CEng.

The fundamental problem concerns the teaching of mathematics in schools, coupled with the limited number of subjects taken at A-level (but not for Scottish Highers) in English schools, since it is here that career decisions are made. It is a scandal that the shortage of competent and enthusiastic mathematics teachers in schools has been apparent yet permitted to deteriorate further, over 50 years, from the time that good mathematics graduates began to find satisfying posts in industry beyond engineering. The situation was exacerbated by the fad for modern mathematics, which taught a limited element of symbolic logic relevant to computer methods but which abandoned the teaching of the means for solving problems. The Royal Society, for example, drew attention in 1998 to the changes in curriculum undertaken without adequate consultation. In consequence, applied mathematics has suffered disproportionately; applied mathematics might be expected to attract the greatest interest from budding engineers in its evident use in explaining and solving practical features of the real world. The Cockcroft Report (Cockcroft 1982) remains relevant, endorsed by the Secretary of State for Education of the time in the context, as ever, of 'the need to restrain public expenditure', without regard for the vastly greater cost to the nation that arises from inadequate numeracy. The Cockcroft core identified the shape of the curriculum for A-level mathematics, endorsed by scientists and engineers of the Royal Society. The needs then, further stated by the Engineering Council (Sutherland and

Pozzi 1995) remain, for trigonometry, algebra and calculus as the basic elements of analysis required by engineers. It merits emphasis that the scientific basis for engineering depends upon calculus, without which experimental evidence could not be expressed in terms of physical laws. The understanding of physical laws is necessary, in consequence, in order to transfer experience from one situation to another similar but different one. Civil engineers need a spatial sense; they benefit particularly, in consequence, from trigonometry and Euclidean geometry (which seems for no valid reason to have sunk without trace).

The school's problem is therefore two-fold: first, that there are inadequate numbers of competent, enthusiastic teachers; second, that the current syllabus is unsatisfactory for those who are going to apply mathematics in their subsequent careers. Enthusiasm is perhaps the most necessary feature, so that the teaching of mathematics may, from the outset and continuously, encourage a sense of 'wonder and curiosity'. In 2003, the Advisory Committee on Mathematics, set up by the Royal Society and the Joint Mathematical Council, has concluded that a National Academy of Teachers of Mathematics is required to tackle the crisis. It is interesting to learn that a similar problem in France has been successfully improved by the same stratagem, so that it is reported that, in consequence, more than adequate numbers of mathematics teachers now exist. The special irony of the protracted inaction to solve the problem is that this has accompanied a period of vigorous growth of demand for mathematicians in finance, commerce and industry. Possibly retired engineers and scientists might help to relieve this serious problem. Immediate economic returns on cost are virtually, therefore, guaranteed.

From time to time engineers will aver that they never use the mathematics they have been taught. Ask a few questions on, say, acceptance tests for materials or the expectation of sound reduction with respect to distance from the source, and it is immediately apparent that they retain basic numeracy, absorbed without ascription into their cognitive abilities. Many engineers resemble in this respect Molière's Monsieur Jourdain who remarked on his discovery that he had been talking prose all his life! Their general expectations of cause and effect are largely based on, possibly only partially remembered, physical phenomena demonstrable by applied mathematics. Fundamentally, an innumerate engineer at the present day is an anomalous contradiction and a positive danger. There is loose talk of engineers working in groups within which there will always be some to compensate others. If the basic level of understanding does not allow competent communication, this is clearly a liability. In any event, from time to time even the most cosseted engineer finds himself required to make personal judgements without others to turn to.

The engineer requires a facility for mathematics for three distinct purposes:

1. for the solution of problems,
2. to develop the capacity for logical thought,
3. to communicate with other engineers and scientists.

At university, the old drawing-office, where some investigated the intersection of surfaces while others meticulously detailed the elements of a riveted structure, is now the studio. This is where the elements of design may be taught and practised. This is also where the engineer learns that creative thinking in design is promoted by the sketch and the simple analysis before a sufficiently developed scheme may be tested numerically; judgement may then be applied in preparation for the next stage of complexity.

The operation of systems lies at the heart of all aspects of engineering. The work of the engineer is subject to hierarchical systems, to developing the relationships between systems in controlling the development of a project, its design, external relationships and, particularly, the investigation and control of risk. Not only is it necessary to understand the operation of the processes of systems (Blockley and Godfrey 2000) but most particularly the ability to communicate, by word and calculation, with those who participate in the operation of the system and with the adjacent systems and sub-systems. Risk has no respect for engineering discipline; this points to the essential need to be able to communicate with those of other disciplines and of different backgrounds.

The student needs to learn about decision-making in engineering, once again applying the notion of the system, in trading-off between the several related factors, including environmental and social concerns. The study of the history of engineering provides a perspective for the understanding of principles in considering why, in the conditions of understanding of the time, particular decisions were made. It provides a context in which to study past successes and failures, pointing to lessons about the 'system' and on acceptable limits of extrapolation of proven behaviour.

The pressure on education to turn out engineering graduates familiar with the more rounded features of the engineer's work is tending to obscure the primary need for those who can think for themselves and work from first – or at least Newtonian – principles. The tools of computer-aided design and analysis in no way supersede the duty of the engineer to establish the criteria for success of a design and in determining that these are satisfied. The ability to sketch and to undertake simple preliminary design calculations thus remain essential stages in achieving these capabilities. Teaching that introduces systems at an early stage can combine the need for logical thinking with the elementary tools of analysis, the formulation of the problem often requiring more fundamental thought than its solution. Much of the supporting technology is then seen early in a university course as a response to a need to know, as opposed to a welter of material to be mastered for examination – and then forgotten. The mistaken motives of concealing differences between IEng and CEng (Chapter 3) appeared deliberately to obscure the importance of the elements of original thinking implied by the profile of the CEng. This may not be a universal requirement but it is certainly a requisite for tomorrow's leaders in all aspects of engineering and its applications.

Different engineers are motivated in different ways. An object of education must be to build upon motivation, whether this be analytical skills for the most numerate or the practical capabilities of engineering design for those best able to project their interests into engineering artefacts. Different university courses should project specific forms of motivation to attract students of types to respond positively to the course. In all aspects of an engineer's education, there needs to be a balance between the provision of 'tools', predominantly through mathematics and physics, and the appreciation of the manner in which such tools may be used. The tool-kit, especially the mathematical tool-kit (Webb and Ashwell 1959), will be under-used and under-valued if provided too far ahead of the demonstration of its use in problem-solving and the explanation of phenomena. It is doubtless true that engineers will approach mathematics with greater enthusiasm once they realise the dependence on mathematics of practical problems they wish to solve, i.e. learning through 'pull'.

The scope for a civil engineering graduate has broadened so markedly in a generation that it is no longer relevant to attempt to predict a career path as a development of responsibility within a chosen application or section of the industry. A common theme remains, however, of the development of understanding of the relationship of the work of the engineer to others, essentially the exploration of the system. The engineer will expect to be exposed to increasingly unfamiliar aspects, needing to understand the nature of their interests and the means for communicating effectively with them. From this ability arises the notion of the 'spider diagram' of Fig. 5.3 (it does not always possess eight legs) illustrating the levels of understanding necessary for the different members of a project design development, a very general description of the manner in which engineers work. It is valuable for an engineer working within one aspect of a system to have experience of other aspects in order more readily to respect insights from different directions. Project management is an excellent training for many other occupations, within and beyond the traditional boundaries of civil engineering.

In Britain, the legacy of the anti-industrial culture of the 1960s obscured the obvious fact that the luxury of an environmental conscience is only affordable on the support offered by a successful industrial base. There remains a cultural problem, built upon the tradition of classical education of an elite, despising numeracy, although it is interesting to note that, at Cambridge at least, in the early nineteenth century the undergraduate had first to sit a paper in mathematics. The predominant problem, however, is that school mathematics is taught by too many without enthusiasm for, or in many instances qualification in, the subject – so the student will naturally find this ennui infectious.

The current attitude of the ICE has been well summarised by its Director General in relation to the acceptance of a degree course by the Joint Board of Moderators as, 'This would be both to understand fully the science that underpins the art of engineering and to develop the necessary skills in logical thinking that maths helps to bring'. The essential element that mathematics brings to the understanding of scientific concepts is implicit in this statement. The objective of widening the field of entry to the profession of engineering should not detract from the broader general good, as a result of implanting interest and understanding of science at an early age.

7.6 The last days of pragmatism

It is remarkable, at the present day, to find chartered engineers who continue to find pride in expressing contempt for the theoretical basis of their profession. As demonstrated by Chapter 1, this is a survival from the British (more particularly the English) pragmatical approach, which appeared to be supported by the low standard of education of many of the pioneering engineers of the nineteenth century. However, the lack of theoretical background of the early engineers was compensated by the entrepreneurial spirit of the time and the opportunities allowed to them to experiment and learn from failure. They were intelligent men who learned from experience, which remains the best teacher, provided that concepts are understood so that learning may be generalised and applied with confidence to similar – but never identical – circumstances. With early education in engineering aimed at the technical school level, the continuity between science and engineering was eroded. The Age of Enlightenment, roughly through the eighteenth century (Porter 2000), had developed an amateurish approach to the useful arts, including engineering, and we continue to suffer from the popular delusion that this view engenders, in supposing that engineering concerns the technical detail, not the underlying concept.

It is beyond denial that the survival of the pragmatic or empirical approach has created barriers to innovation, particularly of construction techniques, as against the design of works for which a coupling between theory and practice has been more generally appreciated and demonstrated. Reluctance for investment, discussed further in Chapter 9, enhances this failure. Britain has paid heavily in importing techniques, processes, special plant and specialist operators mostly developed in the continent of Europe and in Japan, where no such hang-up on background theory has occurred. The present overdue vision and practice of purposeful continuity between design and construction has been largely inhibited by the interference of the lawyer in the relationships between the Client as Employer, the Designer or Engineer and the Contractor. A new age of enlightenment begins to show positive results with innovation both strategic (e.g. by partnering and application of the Observational Method) and tactical, in a return to co-operative working on site.

We are all affected in our philosophies by the attitudes around us – our nurture. I have found much interest in working with engineers from other European countries, especially France, who take a more theoretical, Cartesian (see Section 1.5), approach than I. Whereas I tended to express a view on the optimal solution to a problem once this was set out, I found that they would approach the problem more methodically, eliminating the alternatives on logical grounds, arriving eventually to a conclusion that might be fairly close to my intuitive solution. What is valuable in this more measured approach is that the reasons for dismissing alternatives are discussed and recorded, in the light of knowledge available at the time. This is a useful discipline, which I have learned to respect. I have endeavoured to instil into engineers who work with me and share my inborn pragmatism, the essence of this approach. The virtue of such a method is that, when or where circumstances change, new information arises or hypotheses are shown to be

The engineer for the twenty-first century

false, the decisions may be reviewed in this context and possibly varied. The changes may be factual or they may reflect people and their opinions. It is otherwise often difficult to establish what, if any, variations may be required. The change may in consequence be glossed over or the entire concept unnecessarily reconsidered. The approach I am advocating is, in fact, the natural consequence of operating a well-found system. Records of the processes of decision-making, related to the information available at the time, are also useful in the rebuttal of lawyers whose post-event wisdom allows them to claim how obvious to them is an error of judgement taken at an earlier point in time.

8
Channel Tunnel

And hence to France we shall convey you safe,
And bring you back charming the narrow seas
To give you gentle pass (Henry V, Act II, prologue ll 37–40)

8.1 The revival of interest from 1958

My career as a civil engineer has been so punctuated by periods concerning the Channel Tunnel that I feel entitled to devote a chapter on episodical impressions, but not to repeat the recorded features of the history of the project, the subject of so many books, papers and official documents. My object is to give a more personal view, correcting, in so doing, a few of the accepted accounts and including some of those episodes that escape a more serious history or learned paper. My direct involvement, in many different capacities, occupied a period of 40 years. In 1958–9, I was the leader of a small British team studying the options in co-operation with a French team under Marcel Rama, who became a good friend. The work was directed and led by the Channel Tunnel Study Group (CTSG), an invention of a direct descendent of the Suez Canal Company. René Malcor was the Delegate, Sir Harold Harding serving as his deputy, at least in name; in fact they were largely complementary in talent and experience, with a high mutual respect, a major factor in the success of this brief operation undertaken on a shoestring (Bruckshaw *et al.* 1961). Much use was made of historical evidence, also using geological data from two marine boreholes and by applying the relatively new forms of geophysical prospection. Ancient samples obtained from the seabed, stored by the Société Nationale des Chemins de Fer (SNCF) since the 1870s – it was said by some in the parcels office of the Gare du Nord – were examined for their micro-fossils in order to map, in stratigraphical sequence, the exposures of the chalk. It is not without relevance to observe that a sequence of several metres, 'missing' from the previously understood transition between Gault clay and Lower

Chalk, had been disclosed by a borehole sunk for the work at Folkestone Warren in 1949 described in Section 6.5. The major part of the work of 1958–60 was related to a bored tunnel with associated terminals, strongly resembling the principles of the scheme eventually constructed. It seemed highly improbable that an alternative would be found more economically and environmentally attractive, but nevertheless an attempt was made to prepare roughly comparable costs for other schemes of immersed tunnels, bridges and hybrids. The CTSG appeared ready to give some benefit of doubt to the optimistic aspirations of the proponents of these alternatives in order to benefit from the associated commercial lobbying. Traffic estimates were also made of a fairly speculative nature.

A period followed in which the two Governments made their own assessments of viability, leading to a more extended investigation for a bored tunnel option in 1964–5, for which I was appointed Project Director, responsible to three British and two French firms of consultants, through a Chairman, Monsieur A. Grange (last French Chief Engineer for the Suez Canal). There were two Chief Engineers for the investigations, the British, Morrison, being generally in charge of logistics, the French, Delooz, responsible for recording and interpreting data. This operation provided a most remarkable example of the unnecessary costs resulting from over-detailed but arms-length control by Government. The work was conducted overall by the CTSG, while managed in other respects, organised and interpreted by the Consultants' team, financed by the two Governments, minutely scrutinised through the monthly meetings of the inquisitorial Commission de Surveillance (which never found an English name), whose number of participants was in the 30s. For the first such meeting, the principal players, Malcor, Harding and representatives of the engineers, were asked to be in attendance. We waited for much of the day, were not given any report on proceedings, nor even told when the Commission adjourned for lunch. We received no explanation for this gross discourtesy. However, remorse must have set in, since subsequently we were treated as fellow creatures, including one memorable meeting, well lubricated, at a Rothschild chateau to the North of Paris, evidently a spoil of war or revolution of the SNCF.

An administrative engineer from the DoT was appointed, presumably to ensure that the team of engineers based at Dover Castle (his franchise did not, so far as I recall, extend to our sub-office in Calais) were not engaged in moonlighting activities; he did not become involved in the planning or execution of the work. Our French colleagues appreciated my appellation as 'Le phantôme du château'. A totally unbalanced amount of time, cost and effort had to be diverted to prepare detailed accounts for each meeting of the Commission. For example, we only obtained an imprest account after I insisted on presenting a statement for one toilet roll (in I think 35 copies). The Commission would not allow appointment of additional clerical staff so the over-pressed engineers had to be diverted to such futile tasks.

Accounts had to be kept in sterling, but this was before the days of decimal currency and computerised accounting. Expenditure in shillings and pence needed, therefore, to be converted into decimals of a pound. On one occasion, 3d had been entered as £0.025 instead of £0.0125. I was told that this error (for an unimportant monthly interim statement with no bearing on final liability) had entered the SNCF accounts and needed to be corrected. I therefore apologised profusely at the next meeting of the Commission and offered a bright new 3d piece to the senior SNCF representative. The Commission made no positive contribution to our work, had no influence on the standards of control of cost imposed by the engineers, but added very considerably to the cost and delay of the work. The impression of the early days was that each set of Government representatives was attempting to outdo the other in terms of aggravation. Our French colleagues were in some awe of their side, which made for internal problems, although the preservation of a sense of humour helped towards improved mutual trust as the work progressed. I recall the shock caused to my French colleagues by my reply to a question by telephone from a British member of the Commission that, from France, would have expected and received immediate response. I indicated that the answer would probably appear in tomorrow's daily report – and, incidentally, what was the Test score!

Initially, vessels needed to be hired for the drilling and survey work at sea. Government required all such contracts to pass through their vetting process, undertaken in so leisurely a manner that opportunities were lost, at a time when charter costs were rising as prospection for gas was developing in the North Sea. Much of the summer of 1964 was dissipated in this way – the last straw being a visit by the Minister of Transport which required the drilling vessel to remain in Dover harbour on the first – and last for a while – calm day, for an operation highly dependent on the weather. The Commission began to lose its collective nerve as winter days and weather contributed to mounting costs and low output, with considerable breakage of steel drill-hole casings.

A 'Cartesian' theory was advanced that attributed this phenomenon erroneously to failure as a strut; this was impossible for practical reasons, so I refused to present a report based on this theory. In fact, it is virtually certain that pressures to continue drilling as the weather deteriorated were the primary cause. As the upper part of the telescopic guide-tube was being lowered to its 'parked' position, for later recovery, current and wave loading in heavy seas would cause high stresses, towards the level of the sea floor, in the lower guide casing as it behaved as a cantilever. During full extension in use for drilling, the cantilever would be propped at the level of the drilling stage, reducing the maximum bending stress for the cantilevered casing by 60–80%.

At one period, the Consultants appeared to be threatened by abandonment by the Commission and simultaneously by arbitration by the Contractors, on account of pressures from France to refuse down-time payments, as specifically provided through the Contract. This was a time for resolution and a steady nerve, dependent on retaining trust with the Client and the Contractors, with a stalwart example set by Sir Harold Harding in particular. Work gained momentum in the spring, with a considerable fleet of ships, two drilling vessels and two self-elevating drilling platforms, GEM 111 and *Neptune*, the latter as it undertook a brief working-up period on its way to the North Sea. *Neptune* was a French rig and there was, I knew, some view among the British engineers that they were denied access to technical information about the rig. There was a pervasive degree of tension between the French and the British; one of my main functions was to resolve such problems of technical approach and protocol, which usually arose from failures in communication. One morning, I was paying my usual first call on the French Chief Engineer, Delooz. A principal engineer, Ken Cross, appeared with the information that a visitor to the castle was asking to see the drawings of *Neptune*. Before any further explanation, I expressed the view that this would be for a member of the technical press, knowing that such a visit was taking place. Delooz produced the drawings with impressive alacrity, and I thought no more of the incident. The British contingent mostly repaired for a sandwich at the Dover Castle, the pub at the foot of the cliff below the castle. On this occasion I found much hilarity. Did I know the real purpose of the visitor's interest? I expressed interest. The visitor was, in fact, a young subaltern from the Royal Engineers on an Initiative Course, with the instruction that he was to prepare a scheme to destroy the platform drilling off Dover. Ken, as a Territorial Army (TA) officer himself, with interests in explosives, had naturally been co-operative but had not expected the help from my innocent intervention. Only subsequently, during a visit by my friend, Rama, did I tell the story to him and Delooz; the former was much amused.

Governmental interest in the project revived in 1970, driven as usual in such projects, as for example, *Concorde*, by political rather than practical motives. In each instance, membership of the EEC was the driver. I have described (Muir Wood 1991) the first practical attempt since 1881 to start work on a tunnel in 1973, in a pattern strange even for the bizarre history of the project. The works were to be built privately, funded by Governments, who would take over operation on completion, the operating authority only to be appointed towards the end of the project. France, as ever, was sufficiently enthusiastic for the project that little notice was given to the contortions of the British, provided a link resulted. The ostensible reason for aborting the project in 1975 was the British Government's late discovery of the additional costs in an adequate high-speed link to London, although provision for this had been made in the CTSG's 1959 Report. At the moment of abandonment, there were technical problems on the French side with their 'descenderie' as this section of the inclined pilot tunnel passed beyond the reach of ground treatment from a temporary jetty extended from the shore. It was as well that the plug was pulled on the project before the in-built contradictions caused more costly conflagration. A Committee, headed by Sir Alec Cairncross, had been appointed in 1974 to advise Government on the economics of the project. Abandonment preceded completion of this task, prompting this laconic observation in para 5.1.1 of the Report

> *We have been conscious, as our work proceeded, that everything seemed to be happening in the wrong order. As our report bears witness we have been occupied in considering how a decision should be taken about the Channel Tunnel after a decision had been taken. Studies on which we were asked to advise were too far advanced to be much influenced by any advice we could offer. And so on. Where the logical sequence might seem to be from the appointment of an advisory group to consultants' studies and so to a decision to build, the actual sequence left us at the end of the line. So far as the particular project we were asked to advise upon is concerned, it has been abandoned and our comments come too late to affect it. They are of interest only in relation to the future.*

In a subsequent conversation with Sir Alec, I told him that this 'looking glass' ordering of events appeared endemic to the Governmental handling of the project. There is yet too little indication that politicians are prepared to accept sound, informed advice before launching each new, costly, possibly desirable, venture in self-destruct mode. Baroness O'Neill's 2002 Reith Lecture on Trust (O'Neill 2002) should be compulsory reading.

In 1981 I was appointed as a specialist advisor to the Parliamentary Transport Committee as they discussed the possible options for revived interest in the project. The railway authority was keen for a privately funded project, for rail traffic only, to be subsequently developed as a second stage as demand arose, for transporting road vehicles on trains. The notion was that the Government would finance the difference in cost of a larger tunnel for this latter purpose. My recollection of the time is that this was a ploy to maintain some political interest – but not too much. A feature of this time was an extraordinary bulky report by Anglo–French bankers on the possible means for financing a revived project. Removing the city-speak jargon and the garniture, the simple message could be expressed in a few pages, namely how to turn investment in the relatively risky construction phase into steadier share-holding in the operational phase. No new objective cost comparison had been made since our admittedly over-simple estimates of 1959; the Government substitute for an updated estimate seemed to be to add another arbitrary contingency sum to an existing one. When, in 1985, a competition was launched for a crossing privately financed and operated, no constraint was placed on the mode. Once again, I was advising the House of Commons Transport Committee, who interviewed each of the proponents of the several schemes. Of nine submitted, four were short-listed:

- Channel Expressway. This started as a pair of tunnels to be used alternately by road and rail traffic, manifestly absurd. It later metamorphosed into a pair of road tunnels and a separate rail tunnel. The estimate of cost represented a small fraction of a feasible value. This appeared to be a pre-emptive bid by British Ferries (under an American Chairman, James Sherwood) which could have found good reason to abandon the project early on, by which time others would be reluctant to declare resumption in interest for another round.
- Eurobridge. This took the state-of-art of long-span suspension bridges into a new dimension, with 4-km spans of an elliptical tube for 12 traffic lanes supported by parafil (Kevlar) cable.
- Euroroute. This 'hybrid' scheme would have a bored rail tunnel separate from a road which crossed shallow water by bridge before plunging to an immersed tunnel by way of helical ramps on artificial islands. The Channel is not a suitable location for an immersed

Civil engineering in context

tunnel on account of seabed gradients, high currents and mobile sand-banks.
- Eurotunnel. The scheme for bored tunnels with a loop system for conveying road vehicles between terminals, following much the system described in the 1960 Report of the CTSG (Bruckshaw et al. 1961).

Having completed my advice to the Parliamentary Committee and no longer bound by confidentiality, I was invited by the technical press to comment on the Channel Expressway proposal, which I did in forthright terms. This immediately prompted a verbal message of thanks from a member of the technical team advising Government, who was still bound by rules of confidentiality, who had found that politics appeared to override any tedious issues such as practicality. Cheapness appeared an attractive virtue, however insubstantial.

Eurobridge would have required considerable development studies. Chapter 6 (Section 6.7.3) mentions a proposal for a demountable footbridge for the Henley-on-Thames Regatta, which might well have served such a purpose on a convenient scale. There might also have been doubts in acceptance of a road bridge on wider grounds of transport policy. I believe that Euroroute would have faced strong technical problems, not directly from the obstruction to shipping channels but because the substantial islands in the strong currents could have remobilised the large shingle banks across shipping lanes.

8.2 The tunnel becomes a reality

My most recent direct association with the Channel Tunnel occurred between 1987 and 1998, as a member of the Dispute Panel for the Project – given a rather more imposing title in French of *Comité des Experts*. Our Chairman was Philippe Malinvaud, astute and *sympathique* Professor of Commercial Law of the University of Paris. The four engineer members comprised Gilbert Dreyfus, whose last project had been the development of Charles de Gaulle Airport; Jean Gautheron, with much experience of tunnelling for the hydro-electric works for Electricité de France; Michael Little of Kenchington Little, consulting engineers, and myself. The original intention had been that two engineers with the Chairman would be members of the Panel appointed to consider each dispute, two others as alternates. It soon became clear that the complexity of the relationships, the issues and the Contract justified all five taking a part in each decision. Effectively, the legal setting for our work implied that the engineers served as technical assessors to the decisions, based on legal construction, by our Chairman. There was considerable discussion in the drafting process and the Chairman always achieved unanimity in our decisions. The Contract was partially derived from FIDIC but introduced a number of complicating features. The term 'optimisation' is widely used but where one party, the contractors TML (Trans-Manche Link) are concerned with capital cost and another Party, ET (Eurotunnel), with operation, the determination of an appropriate balance is fraught with problems. TML would need to depend on ET in any estimate of capitalised operational costs. Optimisation was yet more complicated for the Channel Tunnel since the capital costs comprised, in simplified terms, a lump sum for the terminals, a Target Contract for the tunnels, and a provisional sum for the rolling stock. Furthermore, optimisation was stated as including provisions for comfort, safety, speed as well as cost which turns optimisation into a complicated trade-off with multiple dimensions, which could never be expressed in simple commercial terms. Was the original estimate of cost (effectively by TML before ET was invented) one to which TML could be kept, or was it subject to variation as the work was changed? To add to the initial problems, the Safety Commission, a vital body in establishing standards, was only set in action a year after award of the Contract.

Bankers are their own worst enemies in relation to their attempts to control costs of major projects. The urge for quick results is directly contrary to the essential feature of determining beforehand what is to be done, what are the criteria for acceptability, internally and by external agents, and what are the uncertainties that may lead to risk if not managed. As an example of failure to think through the contractual basis, the Target Contract for the tunnels was based on a provisional diameter, so that the target value was likely to change on account of uncertainty as to the basis of the price. The optimal size of tunnel depended on aerodynamics which, in turn, related to the design speed of traffic, to which also was related the power of the locomotives and the resulting heat generation. The complexity of such relationships for a vital basic dimension illustrates the impossibility of undertaking project optimisation as a simple linear process. The critical path for a construction programme will be delayed if basic decisions in optimisation are not taken before awarding a construction contract. Essentially, the project definition process had not been addressed as a system nor had risk been seriously explored, which would have pointed very clearly to the desirable features of decision-making procedures. This would have

Civil engineering in context

indicated an optimal order in reaching decisions, to reduce iterations, which would not relate well to the order of decisions needed for construction. A period of 18 months or more should have been allowed for such deliberations prior to the start of construction, during which time proposals would be tested against the requirements of external agencies.

The Panel adjudicated on 15 references, some others withdrawn as a result of subsequent internal settlement between ET and TML. It is not possible to indicate the aggregate sum in question but it probably affected at least 25% of the total cost, i.e. reckoned in billions of pounds sterling. The contract having been assembled on the basis of commercial and legal terms, many of the disputes were concerned with legal interpretation, wholly unnecessary if the basis had been better related to the actual decision processes and working practices of those to be directly engaged in the project. Decisions had to conform to common principles of the law of France and the law of England, a feature which sounds unexceptionable until discrepancies between them are encountered. For example, in France there appears to be significance in the intentions of those who draft a document, whereas in Britain this is not a feature of interpretation. Reverberating through the disputes came the phrase *Pacta sunt servanda* – the contract must be honoured – an expression which prompts the observation: 'so what?'

It had been suggested that the civil engineers of the Panel would find problems with the many aspects of engineering that would be involved. We indicated that, while we were permitted to have access to specialist assessors, those who addressed us on technical issues should be able to explain, and subsequently argue, these issues in terms we could follow – and such was our experience. One of the most remarkable episodes concerned the introduction of 'claims engineers' on the part of ET with no prior experience of the project. The worldly engineers of the panel were unimpressed by their testimony.

At one time it appeared that the intention for constructive cooperation between Tony Ridley as Chief Executive of ET and Philippe Essig (a past Chairman of SNCF) as Chairman of TML might lead to a resolution of the most expensive problems. Their shared style evidently clashed with the deliberately confrontational philosophy which, by that time, had deeply etched the relationships between the principal parties, however cordial were the personal relationships between engineers. My strong recollection remains of principal figures within the project whose enthusiasm for beating compliance out of the other side compromised their stated determination towards a successful project.

8.3 Procedural defects of decision-making

Others have commented on the extraordinary manner in which the project was conceived, after so many years of deliberation by politicians. There was a brief period of seven months (March to October 1985) for the preparation of offers for construction, with remarkably little thought for the essential interests of the operator. The 100 or so banks, who arranged finance and drew up bonds for investment, appeared beset by the single objective of most rapid completion to earn the earliest return, without thought for feasibility of procedures. Neglecting inflation, which in general terms may be considered as neutral, the project was 72% over budget and two years late.

With a gestation period of nearly 30 years since interest revived in 1958, Government showed a lack of understanding of the breadth of their role in such a major project, regardless of whether it was ultimately to be a private venture, a Governmental investment or something in between. The major political concentration was on the Treaty between Britain and France, covering the political, diplomatic, legal issues. In addition, a small group under Brigadier John Constant was established in 1967 to look at the technical features, building upon the previous studies. Nobody in Government seems to have set down even the most rudimentary rules for procurement, which would be fundamental for success. The future operator should have been the first concern, the body which would then constitute the client for construction, whether private, public or hybrid. The phase of project definition would then proceed, publicly financed in the first instance – to be recovered if the project were deemed viable, taking account from the public point of view of the external benefits. At a very early date, and well before commitment had been made to cost, the Safety Commission would be established to set out a clear set of rules, related to national practices, coupled with an operational role comparable to the principles of the Railway Inspectorate. Standards of tolerable risk would conform to the principles of ALARP (as low as reasonably practical), largely dependent on risk analysis undertaken as an essential part of project definition.

An illustrative design would be prepared as a basis of estimating costs and performance, with matters affecting the environment, planning regulations and safety agreed prior to further commitment. Basic optimisation studies would be undertaken relating performance standards to operation. Precise optimisation is not to be expected since there are too many unknowns; the shape of a curve relating tariffs to profitability is fairly flat (Fig. 8.1). The important point is that a

Civil engineering in context

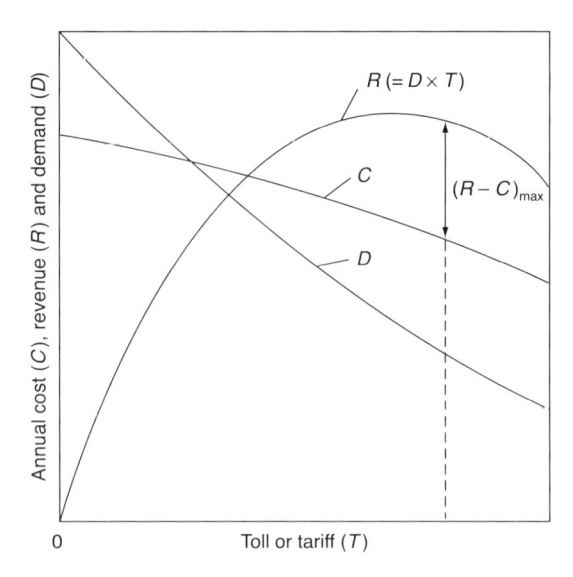

Fig. 8.1 Traffic, tolls and profitability

single body should have the power and responsibility to set out the performance criteria.

A team would then be assembled that was best qualified to undertake detailed design and construction, appointed by competition (not price) in a manner to give high likelihood of achieving good value for money. Characteristics would include total commitment to the project, capability of innovative engineering, combined with a track record and familiarity with management techniques appropriate for such a complex major project. The consequence might develop to a full 'partnering' arrangement, or represent enlightened operation of a FIDIC-type contract. The essential performance-related aspects of design would remain with the Client as Operator, advised by his own consultants 'embedded' in the organisation. Detailed design to achieve the required features would remain with the Contractor(s), with a strong incentive to perform and propose superior means for achieving the standards required. Any bonus payable would depend on the overall achievement of project targets relating to the date of start as a commercial concern, hence encouraging the greatest degree of co-operation between individual contracting elements. Where the contractor's undertakings relate to lump sum or measured elements, uncertainties over which he has no control would be the subject of Reference Conditions (see Section 4.5).

I know personally a number of those in Government, banking and insurance, who have played a part in the organisation of the project, many of these prominent in the Major Projects Association. Management and risk were important aspects of discussion, so it remains particularly remarkable that advice from engineers, with direct experience of the potential problems of a project of this nature, should have been so totally ignored. There is nothing unexpected in the experience of the Channel Tunnel, only the surprising unexpectedness to those who took upon themselves the disastrous decisions about procurement. Technically, the engineering has been good throughout the planning, design and construction, although a certain amount of modification was needed as a result of late setting of standards. The memories of the public concerning the overrun in costs will unfortunately outlive the recollections of engineering innovation and achievement.

8.4 Comparison with the Øresund project

I have elsewhere (Muir Wood 2000) compared the administration of the Channel Tunnel project with that of the Øresund Link between Denmark and Sweden. The timing of this latter project (1995–2000) could take benefit of the experiences of the Storebaelt project between the islands of Fyn and Sjaelland of Denmark (1988–96) and of the Channel Tunnel. Each of three main contracts for the bridge, the dredging and reclamation and for the immersed tunnel for the Øresund Link had a Disputes Advisory Board (DAB) composed of three practising engineers, with experience through the areas of types of work. The project had essentially been conceived by engineers, who determined the procurement process as well as the contracts for construction. The arrangements were exemplary (Reed 1999). Time and budget were respected, with final payments coinciding with the opening of the project to road and rail traffic. There was not a lawyer or management consultant in sight. An interesting feature was that the staff of the Consulting Engineers, advising the Client and who prepared the Illustrative Design, were later absorbed into the Client organisation (Øresundskonsortiet) to ensure full co-ordination between design and operating requirements; the DAB thus inherited some of the objective responsibilities of the Engineer. In practice, the DAB discussed potential problems openly with the parties, made every endeavour to enhance the excellent relationships between them, and encouraged internal resolution of differences. An accountant might well question the value of the DABs; they had no formal claim to advise upon. I believe that the existence of the DABs, with their obvious concern with engineering principles and practice, helped to nurture constructive solutions to each problem. This would not have been effective without a set of Contracts based on a fair allocation of risk, with a liberal usage of Reference Conditions. The Illustrative Design had been used to probe the several potential problems with external agencies, particularly the severe environmental requirements of the locality. This procedure was of great value to the smooth progress of the work and to the absence of major confrontation with green issues.

8.5 Early history of the Channel Tunnel

It is interesting that, while the Channel Tunnel project had haunted the corridors of political power and prejudice for so long, controversy remains as to how long. Popular accounts attribute the first positive step to a suggestion by Albert Mathieu to Napoleon during the Peace of Amiens of 1802, the brief period of respite from conflict between Britain and France in the Napoleonic Wars. The most comprehensive account of the early history of the project (Lemoine 1991) from a French viewpoint, however, hints darkly that the discussion between Napoleon and Charles James Fox on this proposal in 1803 is a figment of the imagination of Thomé de Gamond. He first mentioned this discussion in 1857, in the absence of any other witness, with an accompanying sketch of an improbable underwater coach route. Thomé (de Gamond was a later flourish) was the most prolific author of several of the most remarkable proposals of the mid-nineteenth century. To Fox is attributed the observation 'This is one of the great things that we must do together' which might have been designed to provide an historical perspective to Thomé's spectacular proposals, while flattering the pride of Napoleon III. Thomé began to study the bed of the Channel in 1833, made intrepid dives to obtain samples from the bed from 1838. He then joined those who devised the range of tunnels, bridges, immersed and buried tubes, islands, mid-Channel ports and long jetties which punctuated the century. This collection of the impractical ensured that the project was unlikely to be taken too seriously; the degree of positive interest can be directly correlated with the condition of the Entente which drifted between the epithets 'Affreuse' and 'Cordiale' through the period (Muir Wood 1991). The geological linkage between the French Boulonnaise and the English Weald, while viewed as probable, remained to be established, together with the elimination of the postulated major rift fault along mid-Channel.

Meanwhile, the geologist Godwin Austen had predicted (Harris et al. 1996) the existence of the Kent coalfield as an extension of the measures already worked in north-west Europe. Early investigation for the tunnel in the 1870s was thus related to the prospection for coal.

There is a hilarious account of the attempt to stop, in 1882, the Beaumont trial tunnel from Shakespeare Cliff at Dover. The Board of Trade had decreed that this adit should not proceed beyond the coastline. An Inspector was dispatched to the site to confirm that this limit was being honoured (in fact, by that time it was well out to sea). On the occasion of the first visit, there were problems in finding a key to gain access and no surveying instruments were available. At the second visit, the position was established and the tunnel advance was ordered to stop. It was argued, however, that the compressed air to drive the tunnelling machine was the only source of ventilation; its interruption would be dangerous. It was apparently only during his homeward journey that the Inspector appreciated that uncoupling the compressed air line would solve the problem without indefinitely continuing the advance.

The chauvinistic remarks in Parliamentary debates of the period reflect the insular, imperialistic views of the time. The Romantic poets combined with the military to oppose the project. Lord Palmerston, in his well-known riposte, expressed the view that the distance between England and France was already too short! Prince Albert, as one might expect, was a prominent advocate for the tunnel. The Duke of Wellington was a powerful opponent (the shares rose significantly at his death) while he was an enthusiastic supporter of the Thames Tunnel, predominantly with an eye for its contribution in moving troops to the south-east of England in the event of invasion. However, von Moltke, stating the views of a German military strategist, observed that 'the tunnel ought to be opposed because it could not be used to attack England and, in case of conflict with Germany, it would be fatal to the latter'. Mr Arthur Fell MP, a later supporter of the project, considered that the presence of the tunnel pre-1914 would have reduced the period of the First World War by two years.

8.6 Random technical reflections

It had been postulated over the years, proven by the 1958–60 and 1964–5 investigations, that the somewhat more brittle nature of the preferred tunnelling medium, the Chalk Marl, on the French side of the Channel, associated with minor faulting, introduced greater problems of encountering water. In consequence, provisions were made

Civil engineering in context

for operating TBMs from France in closed mode, i.e. with the face pressurised. On the British side, experience with the behaviour of the Chalk Marl under the land encouraged a general expectation that a comparable degree of lack of competence, i.e. the ability to deform and hence close fissures, would be experienced under the sea. In fact, over a length of about 2.7 km, possibly as a result of combined warping, reduced cover and weathering, ground treatment was required to control water inflow. For a future project in comparable conditions, it should be possible, probably by the use of in-situ testing of a type to overcome questions of scale (e.g. the distance between rock joints relative to the diameter of the probe), to avoid such surprises.

The nearest approach to a crisis in the tunnelling resulted from excessive convergence which occurred during excavation of the crown of the cavern for the UK under-sea crossover. The cause was identified as locally high water pressure in a permeable layer of chalk within the Chalk Marl, which needed in consequence to be drained by means of drill-holes. In general, in layered ground of this nature, precautions should always be considered against such a risk within a specified distance of a tunnel, the distance depending on the ground, the geometry and the maximum possible water pressure.

Another issue concerns the practicability of construction over time. From the mid-1880s there was reasonable expectation of geological continuity, without which a nineteenth century under-sea bored tunnel would have been a disaster, so feasibility after that time depended on technology. Available mechanised means were crude and slow; although tunnelling machines had been used for headings from each coast, their diameters were little more than 2 m. Many years were to pass before a machine of 7–8 m diameter was developed. Logistics would have created major problems, also how to cope with water inflows and ventilation. The incremental developments of the early twentieth century would have helped towards solving such problems but progress would have been slow with periodical stoppages to deal with local geological difficulties. By the time of the studies of 1958, the project would certainly have been feasible but assumptions made at that time in relation to the control of water were almost certainly optimistic. The advances in TBM design over the past 20 years have been the most important factors towards economy in construction. Probably, therefore, the tunnel was built, in technological terms, at about the most favourable time; it is only a pity that the administrative mishandling and the failings in procurement strategy have obscured the technical merits of timing.

One feature of construction bears remark. On the British side, the main working site alongside the old Shakespeare colliery, was very cramped, with more space only becoming available as progressive reclamation became possible, using spoil arising from the excavations. In consequence, the organisation of the work depended upon resourceful improvisation, with periodical rearrangement exploiting the available space. On the French side, on the other hand, the main working site at Sangatte near Calais was virtually level and unlimited. Here, the French penchant for planning at a grand scale could be exercised to the full. Tunnelling was all undertaken from *le grand trou*, 55 m in diameter, 65 m deep including a 21 m sump. (The story runs that there was a lapse in applying for planning authority – the hole was explained away as a minor trial pit). This concept allowed the operation to be undertaken from the principal work site as if it were a factory, with high standards of cleanliness and organisation.

An energetic young French engineer had been posted to Dover in a liaison capacity. I suggested to him that he could write an interesting thesis on the different approaches between the two sides, differentiating the features imposed by topography and geology, from those caused by different national approaches. Undoubtedly, the French system of engineering education and training encourages wider vision; the British system, the better use of pragmatic means to overcome awkward problems by improvisation. I suggested, therefore, that he should also speculate on what might have happened if the French had been faced with the British conditions and vice versa. I do not know if he ever tackled the subject.

So, we have our Channel Tunnel. Technically, the engineering was imaginative and good. The procurement issues were poorly planned and executed; it is unfortunate that the public perception is of overruns in cost and time. It is not enough for the engineer to explain that this was not his fault. It is the job of the profession and our institutions to ensure that we are not in future put into this situation by those lacking a penetrating understanding of the Client's interests, which should, as ever, have been the prevailing consideration. Until work was about to start, the Client for the Channel Tunnel did not even exist.

With the prospect of the Channel Tunnel and its link to London, to be completed by 2005, so superior in speed characteristics to any other surface link to the capital, we may yet see two-way commuting between Britain and France, as has followed completion of the Øresund Link between Sweden and Denmark.

9
Ethics and politics

9.1 Introduction

Is the connection between politics and ethics fundamental or superficial? The cynic will have a ready reply; to the engineer, the association is essential. Personal ethics of an engineer is a reflection of the political consequences of engineering decisions. Unless the engineer shares the philistine notions of extremist politicians, he must recognise the social and environmental consequences of engineering solutions to practical problems. In the absence of understanding the mechanism for turning engineering aspirations into policies, the ethical engineer will experience no more than a futile 'feel good' factor. As R. H. Tawney has reminded us in *The Acquisitive Society* (Tawney 1921): 'The difference between industry as it exists today and a profession' is that 'the former is organised for the protection of rights, mainly rights to pecuniary gain' while 'the latter is organised, imperfectly indeed, but nonetheless genuinely, for the performance of duties' (1961 ed. p. 89). The essential feature of the professional, if he is to have influence, is that he must be trusted for objectivity, as well as high professional competence, imagination and acumen. Too many glibly quote Shaw's dismissive epithet: 'All professions are conspiracies against the laity' (*The Doctor's Dilemma*) unaware of the context of a peculiar sample of the medical profession engaged in fashionable practices ('quackery'?) beyond their comprehension. He might equally have had the legal profession in mind. Through the ages, the special privileges the legal profession gives itself have been associated with the notion that lawyers do not always subjugate their own interests to those of their client. With the recent rise of the large firms of commercial lawyers who pay themselves generously to match the income of their clients, these criticisms do not subside, rather are they increasingly voiced from within the legal profession

Ethics and politics

itself. As one with considerable experience as an expert witness of working with lawyers, I find a curious anomaly. The working solicitors are expected to work excessive hours (a recent example is that of an engineer turned solicitor, encountering his responsible Partner on leaving his office at 7.30 p.m., being reminded that he was no longer an engineer) under pressures to maximise (time-scale) earnings, against the rationalisation that this is all in the interest of the client. An objective observer would rapidly conclude that the quality of output would be markedly improved by a less stressful working environment coupled with the aim for the earliest settlement – but of course the profits would suffer!

I understand morals as representing the ability to choose between right and wrong; ethics as the philosophy behind the ability to make such distinction. If we reflect upon generalised views of ethics in the late nineteenth century in Britain, there appeared to be a few accepted principles. These were centred upon religion, on loyalty to an imperial nation, on aversion to debt, to dependence upon a scarcely trammelled capitalist system, and to a general conformity in thinking. By contrast, at the present day, there is much greater diversity in acceptable life-style, in tolerance to running unaffordable personal debt, in hedonistic attitudes devoid of allegiance to any higher authority, even in the definition of what constitutes ethics. Does it, for instance, entail adherence to a faith beyond humanism? How sacrosanct is the environment? Is there an assumed liberty of action, provided it does not injure others? In many respects, the definition of 'utilitarian ethics' becomes simple in so far as it does not imply support for a broader 'good' whose merits need to be separately argued. While at the fundamentalist level, the hostility between religions is more intense, fanned by populist politicians, at another ecumenical level the acceptance of common ground between the world's great religions approaches the mutual regard and tolerance so forcibly destroyed in the fifteenth century in Europe and subsequently elsewhere. It remains a fact that wars between religions have not approached the ferocity of wars between sects of the same faith.

This is not the place to argue whether or not the moral ethos of the society in which today's engineer works is advancing or retreating. It is certainly changing, with a declining respect for authority, accompanied by high popular regard for evanescent celebrity figures who occupy their own amoral society. The engineer can no longer assume that he shares the notions of loyalty and strict confidentiality, for example, with those around him. There is further confusion in the manner in which the term 'ethical' has been hijacked for purposes of financial advantage. Thus,

Civil engineering in context

definitions of 'ethical investment' may be found to be based upon subjective or popularist criteria, so that one such definition would exclude road building, on the false hypothesis that the environmental considerations will always prefer an alternative to building a road. Corporate Social Responsibility is known by the acronym CSR, revealing that the management consultant has been here! The claim to observance is displayed like a logo, and like a logo, the features of observance may be related cynically to the commercial interests of the organisation rather than any deep ethical concern for its virtues. There is even a point scoring system, operated by accountants; high marks depend on understanding the rules of the game and using the right terminology. It remains necessary to explore behind the façade to distinguish the genuine players from the mountebanks. It is far more valuable to have insight into the objective nature of the ethical strategy of an organisation than to a list of deemed 'ethical' features of a questionable basis.

While, therefore, a greater fragmentation of the philosophy behind popular expressions of ethics exists than a century ago, there is a greater agreement that ethics essentially implies responsibility for the well-being of others in the present and in the future, related to the consequential social and environmental dimensions. Such a broad definition leaves liberty for a wide degree of translation into actual personal and collective behaviour. What means are justified by the ends? To what extent can people's views be distorted by propaganda and misrepresentation? Often, the preferred course may depend upon the 'frame', i.e. is a high technology solution to a vital problem in a developing country justified by the immediate severity of the problem to be solved?

These are broad issues which continue to be debated. By focusing more specifically on the issues affected by decisions and actions of engineers, we may reach a high degree of unanimity of the moral criteria to be considered, if less on the subjective question of ethics based on assessment of their relative importance. Armstrong *et al.* (1999) attempt to set engineering ethics into its wider context of simplified philosophical concepts.

9.2 Engineering ethics

The engineer, as any other individual, will have more or less of a moral attitude to his relationships with others, and for responsibility for avoiding harm of different varieties. It is essential that he differentiates between those aspects of ethics which derive from his profession from those that do not. For example, the several shades of pacifism represent honourable moral positions, but have little relationship with engineering, beyond the aspect that engineers of certain disciplines may have special knowledge of the consequences of particular forms of warfare. Engineering ethics implies that the engineer will have special concerns for the consequences of his actions as an engineer. I have declined invitations to memberships of bodies with titles including 'Social Responsibilities of Engineers' or similar, for the specific reason that their manifestos invariably extend beyond the engineering dimension. I find disappointing that those who set up such organisations are unaware that their statement is weakened by failure to observe this logical demarcation. Their professional judgement should be founded on a more secure factual base than the wider moral attitudes, which are subjective, have shades of value and are arguable.

The ethical engineer needs to recognise different stages in relating his concerns to politics. He has special skills to identify the undesirable consequences of an engineering decision affecting society or the environment. He can suggest avoiding actions or compensating policies, in certain respects, so that these concerns may be expressed in money terms, allowing policy decisions to justify the additional cost. Often, however, there is no such simple equation: some ecological degradation, some undesirable consequence will be inflicted on a valuable site or a group of people. The choice of action, the association of benefit for some with cost (in the broadest terms) for others, must be a political decision, doubtless taken within policies laid down at national or wider level. Here, therefore, the engineer is concerned with identification and appropriately specific description of the consequences, so that the political decisions can be taken against the most reliable professional opinion. Various attempts have been made to establish costs for qualitative features, but these are bound to fail. For example, economists have endeavoured to price 'heritage sites' by calculating ('environometrics'?) what the visiting public is prepared to pay, in terms of time, travel and entry costs. Perhaps the most valuable sites, in uniqueness terms, are the least visited, whose value is enhanced by their isolation and tranquillity. 'Environometrics' would condemn these to destruction. This is an example of where the engineer should recognise the capability of other professions to provide advice on aspects of their particular profession, in this instance in terms of comparative quality. We do not look to the accountant to value our historical buildings, but to the art historian, the historically minded architect, the student of local history. We cannot hang on tenaciously to every feature of the past; the skill must be in conserving the best and to make best endeavours to fit what we keep into a modern and changing world.

It is instructive to consider how the scope of the work of the civil engineer has changed over the last 150 years. In the mid-nineteenth century demands for transport, energy, water and drainage were overwhelming, particularly in those countries of rapidly increasing urban population. The rural scene, in which much of such development took place, had matured over the centuries and seemed capable of absorbing development without undue pain. Monetary compensation was usually deemed adequate compensation (at least to the local landowner) for the right to build the necessary works. In these relatively early days of industrial expansion, technological excitement was a partial counter to the loss of rural peace. As development becomes more intense, so does conservation assume increasing importance. It is, therefore, a natural consequence that the duties of an engineer must, in time, become increasingly involved with aspects other than the purely functional objectives. Attitudes need to be judged against the priorities of the times. We need not criticise our predecessors who, against today's criteria, may seem to have been insensitive.

There is one respect in which engineering ethics is too readily aligned to an over-green philosophy. Nuclear warfare is discarded as unacceptable (except possibly by its use, by some, as an ultimate threat). Nuclear power is too readily damned by association, in the recognition – in so far as logic may be applied – of its early function in providing isotopes of value in making warheads. Early reactors were associated with wastes of long life, considered as a lethal threat to posterity, extending beyond a period of extrapolation of today's societies. The first objection may be countered as being one of association; if we deal responsibly with our wastes we are not contributing to the international problems of the threat of nuclear warfare. The second objection may be answered likewise, in that accepted designs of nuclear power do not give rise to nuclides of life beyond that of reliable containment by capsules. This is the basis, for example, of the well-engineered system for nuclear waste disposal in Sweden (Royal Society 1994). There are existing long-lived nuclides in Britain that call for special treatment in disposal, but this is a fact of life unaffected by future energy policy. Transmutation might be part of a future strategy for their disposal. There are many threads to the development of a viable energy policy, combining low-energy consumption with different forms of renewable energy. It is clear that, at the present time, by reluctance to engage in debate on the merits of nuclear power the UK Government is failing to develop an energy strategy which reconciles needs with undertakings to reduce carbon emissions, the consumption of hydrocarbons and features of national security. Engineers should not add to the difficulties of the politician by spurious apprehensions, apparently technically supported through their professional allegiance.

Efficient engineering, in project concept and execution, is essential for ethics to flourish. Efficiency implies transparency between all participants. Efficiency only provides a vehicle, not a guarantee, of ethical success. Ethics also dictates adherence to the law, but legal requirements provide no more than a framework for ethical behaviour.

Ethics clearly relates to risk. An inherent part of the engineering process is to perceive exposures to potential risk and to evolve appropriate strategies. Risk needs to be studied not only in physical terms of the possible developments from uncertainty but also in terms of human behaviour. For example, if a form of contractual relationship sets people to assume responsibilities in an unfamiliar or equivocal manner, the potential risks should be anticipated by actions to ensure these are understood and that appropriate training is provided. As a special feature of such situations, top management may understand the changed scene, operatives may be given a new rule book, but middle management is often resistant to changing habitual practices. I have commented elsewhere (Muir Wood 2000) on the contribution of this factor to the collapse of the Heathrow Express Tunnels in 1994.

It is too easy at the present day, beset by regulation, quality assurance and standard procedures, for the engineer to assume that, these hurdles surmounted, his duties, including their ethical dimension, are achieved. In the early days of motoring, there were no road signs or markings, no procedure at intersections. The judgement of the motorist was the means of limiting accidents. The present proliferation of mandatory and cautionary signs and signals tend to lull the driver into the belief that safety depends solely on their observance. In fact, as much judgement as ever is required, combined with far greater power for inflicting damage; there is also increasing dependence on the assumptions – fortunately usually reliable – as to how other road users will behave. This is comparable to the situation of the engineer. He is helped by having the scope of many of the ethical issues spelt out (see, for example, Section 7.4), but should not rely upon their comprehensiveness. He may even be provided with methodologies for assessing particular issues. He must nevertheless resist the attraction passively

Civil engineering in context

to follow the regulations without, from first principles, identifying the important issues for the particular case. Armstrong *et al.* (1999) provide examples for the several stages of implementing the Mount Pleasant Airport in the Falkland Islands, set against the Rules of Practice of the Institution of Civil Engineers.

The engineer should not assume that other professions with which he will engage will share his objectivity and sense of duty, nor will they necesssarily share a common professional code of conduct. He will need to understand that much of his code is, in fact, specific to his profession. Once again, he has to be careful not to confuse the special insights of his profession with personal views, on which his opinion has a value no greater than any other citizen.

Jonathan Sacks, writing in *The Times*, discusses the shared moral values of a society which should be in the nature of a covenant, embodying basic principles, to be compared with a contract which is specific for the occasion. In much the same manner, the ethics of engineering should be considered as a covenant of fundamental principles, recognising that the details of application will vary with time. The decisions affecting a project will entail 'trading off' between the benefits and costs, the former often enjoyed by a limited number and certainly a different grouping from those who bear the cost, directly or indirectly. In the early stages of development, the environmental effects may be considered as negligible by comparison with the benefits. As the basic needs of society become more adequately satisfied, and as a regard for consequential effects become better articulated, so does the equation change. This may lead to the abandonment of a project, previously seen as meritorious, or to overall environmental protection of excessive cost. Currently, for example, there is a debate as to whether major impoundments of water for hydro-power may make greater contribution to global warming, through methane generation as a result of decaying vegetation, than the alternative forms of generation. Where the environment is being encroached it becomes a wasting asset whose value increases with time; this only adds to the difficulty of expressing environmental values in terms of present value costs.

It is often the engineer, or his scientific counterpart, who first foresees an incipient undesirable side effect. He then has a duty to provide appropriate warning, couched in terms which may help to avoid irresponsible overreaction by the media, but which will alert other engineers to propose appropriate counter-actions to their clients. There is great value in the engineer making known his apprehensions for matters of safety of social or environmental damage before these set off a major

Ethics and politics

cause for concern. The political consequences of an accident often leads to overreaction, with greater costs in squandered resources and disruption than more measured anticipatory measures. In due course, the warning action may be codified into regulation. The engineer consulted on their content should advocate simplicity, absence of complexity requiring legal interpretation and avoidance of over-prescription which may encourage more damaging alternatives. Excessive cost spent on safety or environmental protection has a negative overall effect. John Rimington (Rimington 1993), then Director General of the HSE, puts the matter thus: 'to pay too much attention to avert harm is likely to increase harm in the long run'. When an engineer has reasonable cause to foresee a problem there is a prescribed route of making this known so that it may be answered or acted upon without conflict with the duty of confidentiality (Armstrong *et al.* 1999). This has been titled 'the whistle-blower's charter'. I have experienced a particular problem where signature to the Official Secrets Act has prevented exposure of a spurious piece of research accepted as authentic by my Client. He was evidently not prepared to experience the embarrassment that would result from this admission – and reliance upon the research was leading, in this instance, to substantial unnecessary expenditure of public money. So far as I know there is no ombudsman for resort in this situation.

Where regulations for Environmental Impact Assessment (EIA) are provided by Government Departments there is a reasonable expectation that these will become over-complex. Environmental features of road construction, for example, extend to several, and expanding, bulky box files, with great concern for the calculation of energy inputs, for example, for contributory operations and materials – where many differences between different approaches may be within the margin of error. Meanwhile, the local environmental and social effects, which cannot be so codified, may constitute the most important features, certainly for those most affected.

Development projects affect surrounding areas to varying degrees. For roads, for example, distance limits are provided for the effects of noise, fumes, light and disturbance on various forms of wildlife. There is no such indicator, and indeed none would be valid, of general application to define the degree of degradation of the countryside, which is associated with urban development. In the future, could one, for a specific proposal, make contours of the effects on environmental standards, not only in terms of disturbance but of increased costs of crime and welfare, for example? There is a diffusion gradient of degradation related to many factors of an urban settlement.

Civil engineering in context

9.3 Engineering ethics translated into politics

Engineering ethics assumes a basis of professional competence. Professions are associated with privilege. Parliament enjoys privilege but grudges such privilege to other groupings – apart perhaps for the law, to which many politicians belong – since it seems to derogate from their realm of power. A natural tension therefore exists between politics and the justification for, specifically, engineering ethics.

We need to start by agreeing that engineers and politicians have markedly different characteristics. Engineers tend:

- to be highly focused on the issue at hand,
- to be methodical,
- to be bold in planning for the long term,
- to prefer the technically justifiable to the popularly acceptable,
- to observe an objective ethical code,
- to accept that inevitably they make mistakes from time to time.

Politicians, on the other hand, tend:

- to expect immediate achievements (or at least within the parliamentary term),
- to be dominated by political agenda,
- to be highly influenced by, and manipulative of, popular attitudes,
- to have an appalling record of the upkeep of the national infrastructure,
- to subsume issues of great moment to inter-party abuse,
- to claim, apart from a notably small minority, infallibility.

To many, the greater relative importance to a Minister of a favourable profile in the popular press over attention to state business, portrayed by the satirical TV programme *Yes, Minister*, was no caricature. The standard of debate in the House of Commons is generally low, abusive and displaying far more interest in winning debating points than in addressing issues of the moment objectively. Major legal measures appear to be enacted without any opportunity for the citizen to learn the arguments pro and contra.

Traditionally, the civil engineer was insulated from the politician. He was presented with a problem defined within his terms of reference, allowing consideration of alternative means of achievement, the relative merits assessed by simple application of cost/benefit analysis. The terms of reference provided the political framework, with the engineer only concerned to avoid conflict with the stated assumptions and

Ethics and politics

background. The social and environmental issues were assumed as secondary, and as already comprised within the terms of reference.

It is not for the engineer to attempt to usurp the politician's function to take policy decisions. It is for the engineer to improve the quality of the understanding upon which such decisions are taken and to provide sound advice on the likely consequences. A bad feature of practice a generation ago — and continuing among the unenlightened — was for the Client to appoint several consultants from different professions to work on different aspects of the same project. Questions of risk and of ethical responsibilities tended to become fragmented; in fact, the only one in a position to address such tasks was the Client, and he was normally not competent to do so. Each professional, presented with terms of reference, in whose formulation he had no part, and unaware of the terms of reference of others, might well assume that the ethical issues figured in the responsibilities of others. Too often these issues fell into the gaps.

Today, the engineer will normally be expected to make a multi-faceted assessment of a project, in conjunction with other professions, including the social and environmental features. This holistic design process (Chapter 5) should have identified the most favourable candidate projects (including, possibly, 'do nothing'), exposing most clearly the trade-off between different aspects which may entail a political decision. The engineer who engages in communicating the conclusions to a wide audience needs to understand the art of presentation, a familiarity with the subject, an ability to select the features of significance to his audience, and the use of language which enhances understanding without excessive over-simplification.

These characteristics influence the recruitment and upbringing of the new generation of civil engineers. As the subject area has broadened, so has the background science deepened, providing grounds for redefining the art of civil engineering. The current initiatives by the ICE to widen the range of acceptable disciplines and occupations should be accompanied by a campaign to ensure that all members understand enough about these non-traditional paths of entry to the profession to engage in constructive dialogue. A shared understanding of the unifying scientific basis for all relevant aspects of civil engineering will be a necessary component of this capability. All those identified as professional engineers should undertake to observe the Code of Professional Conduct. Above all, the opportunity should be used to explain that mental strength and agility are as deeply challenged as in any other profession.

The 'new' engineer must not suppose that, for success, he may rely on presentation to mask his technical deficiencies. He may gain support for mastering his subject from Plato's Socratic Dialogue, *Gorgias*, in which the argument is strongly developed for the superiority of the competence of the articulate professional over the rhetorician (today's PR executive or spin-doctor). Are we yet educating and training our engineers in these directions? My perception is that the project fragmentation of the 1980s – politically induced by an administration which advocated commercialism as a religion – has left a legacy of work fragmentation in the office. A high proportion of engineers, who continue to work on cost-limited fractions of a project, see only a limited facet, insufficiently enmeshed with those engaged with other facets of the same project. The interface problem continues and the training to engage with the politician is thereby thwarted. It is essential that those who understand the fundamental technical details of the project are also those – not necessarily contemporaneously – who learn how to present the whole. The mix between specialisation and generalisation remains important. Training needs to be designed to improve depth and breadth of the engineer's capabilities.

An engineer who engages the momentary attention of the politician often concludes that the passing interest has been occasioned by the hope of the politician in finding support for some quite different agenda. The politician is quick on his feet and the engineer has to learn to be correspondingly nimble in returning the discussion to the specific issue. In particular, the engineer has to prepare for the occasion, to think around the issues and to identify the features to emphasise of likely interest to the politician.

There are initiatives designed to assist the politician to understand the trends of science and engineering. These are laudable but their organisation needs to confront the apparent short attention span of the politician. The House of Lords provides opportunity for a few outstanding scientists and engineers to direct their knowledge within the political debate itself, ensuring that these concerns cannot be dismissed as irrelevant to the particular case in point. It will be vital that any reform of this House should recognise the great and increasing value of this contribution.

The greater the degree of communication between the engineer and those capable of influencing public opinion, the less the excuse for the peddling of misconceptions which add to the engineer's problems. The engineer needs to cling to his aim towards balancing the claims between improved facilities and the protection of other interests. There are many who are trained in other disciplines to recognise the negative characteristics of any development but the engineer alone, through his specific professional training, has the duty and the capability to find balanced solutions.

It is the short-termism of politics which engenders many of the problems of the infrastructure, in education, transport, housing,... The legacy of problems from one group of Ministers to its successors engenders excessive attention to acute short-term issues in place of effective policies, which may have formed the manifesto for the electorate. At least in principle, regional government with control over funding (corresponding to the States of the US and Australia, the Provinces of Canada, the Départements of France, the Länder of Germany) allows the immediate issues to be treated locally by one group not hobbled by departmental rivalries, freeing at Federal (i.e. National) level energies for strategic action. Perhaps surprisingly, this additional level of government appears to entail reduced unit cost in relation to our over-centralised arrangements. Co-operation in overcoming the handicaps in poor infrastructure should take precedence over the party political squabbles.

It is probably inevitable that politicians allow excessive influence from the world of finance, which has the readiest access to wealth, albeit largely accumulated by others, and, hence, at least in principle, provides the readiest source of taxation. The engineers need to exert greater efforts to persuade the bankers and insurers that a technical understanding would permit them to take wiser decisions to safeguard the funds of others – and their own. In the esoteric reaches of investment in derivatives and similar secondary features of the financial market (lacking any obvious social merit) specialist areas of applied mathematics have developed. A generation ago, consulting engineers attempted fruitlessly to persuade merchant bankers to direct funds to developing countries into controllable projects, as opposed to loans to national leaders, which were secured, as we were assured at that time, by sovereign risk. Sovereign risk for a bankrupt nation with corrupt government is now too evidently a charade and the funds, with their potential for massive social improvement, have been dissipated. The opportunity has gone, but the experience, associated with arrogance, should not be repeated.

Where notable progress is now being made is in the widening acceptance of the importance of systems in enabling intentions to be executed as achievements. Acceptance of the need for control of the interactive processes and the overriding need for continuity of responsibility are two features which require constant emphasis to the politicians. While, for the engagement between the engineer and the world of money, these principles have most direct application to project finance, there is a wider lesson for all aspects of the execution of functions in the achievement of policy. The excessive costs of PFI, adding up to 25% to the cost of a state-funded project, could be greatly reduced by improved clarity in objectives, serious attention to the features contributing to uncertainty and by demonstrating greater competence in the management of risk. Since the high costs of PFI feed in to the financial world, it is unlikely that advice on its reduction will come from here. Here, once again, is an important role for the influence of the engineer. Those who apply science to the prediction of the magnitude, locations and inter-relationships between the several forms of natural risk are already influential with the major sources of insurance and re-insurance. In all areas of scope for improved decision-making by politicians, the engineer brings a synthesis between the several supporting sciences and the art of their management. The escalation of costs of the Channel Tunnel (Chapter 8) provides one of many examples of the cost of ignoring the systems and risk management of sound engineering. With support from notable recent examples that demonstrate the virtues of a systematic approach (Øresund Link, Channel Tunnel Rail Link), the engineer no longer needs to base his counsel on successes of the past, of the days before the principles were polluted by the excessive commercialisation of the 1980s.

9.4 Personal excursions into the political scene

I have had several opportunities to engage in activities beyond the normal scope of the functions of the civil engineer. I describe below a few of these experiences which have flavoured my understanding of the frontier between the engineer and the politician. Most have related to service on advisory bodies of different varieties.

One curious example of several experiences of public inquiries stands on its own. The extension of the M6 motorway north from Lancaster is a much admired stretch as it climbs over the fells on the way to Penrith, the alignment well adapted to the open country. This length of road was associated with the Kendal spur, a new road to Kendal from the motorway. The selected line for this road would have crossed Levens Park, intersecting a fine avenue designed by Beaumont, one of the earliest examples in Britain of formal landscape design, of the late seventeenth century. Mr Robin Bagot of Levens Hall had assembled a strong team, including landscape architects and garden historians, to oppose this proposal. The obvious course was to find an alternative route avoiding such environmental damage. Initially, there seemed no great difficulty but, once the DoT had rejected the chance to reconsider the decision, as the affair developed so was this position more deeply entrenched, affected by a degree of *amour propre*. Having agreed the unit costs that provided a basis for estimating comparative costs for assessing an alternative route, these were periodically changed. I complained to Robin Bagot that we were being expected to play a form of Chinese whist (where the rules are declared only after the hand is played!). I then learned that, in support of the aesthetics of the proposed scheme, the associated crossing of the River Kent was being displayed and admired in an exhibition for the Royal Fine Art Commission. I inspected the site and the picture. By the addition of a few trees and a heavily disguised causeway, the highly asymmetric layout of a bend in the river, an eroding cliff at the right bank and a low expanse of deposited material at the left bank, had been skilfully obscured by the artist. This was a curious convolution of nature into PR art. In due course responsibility for the project was transferred from London to the North West Road Construction Unit; we were then able to agree that the decision would be based on merit not cost. In due course, the road was rerouted and Mr Bagot recovered his full costs of opposing the scheme, I believe the first to do so.

9.4.1 ACORD

ACORD (Advisory Committee on Research and Development) for the Department of Energy provided an exposure to the potential of alternative forms of energy and processes of energy conversion, including improved efficiencies of traditional sources. A body of this nature is liable to be hi-jacked. Their reports are, at least initially, confidential to the Department; a sense of frustration is shared by its members when a Minister proclaims selected fragments of advice and then prevents the publication of the complete report which could set these quotations in context. The existence of such a body at the present time (accepting that the Department no longer exists) might provide a stronger discipline for energy policy, which currently displays a high 'Micawber factor' – a pathetic reliance on 'something turning up'.

Our discussions on offshore wind and wave energy considered the costs of bringing the energy ashore, coupled with the absence of contribution to peak energy demand from sources dependent on the whims of the weather. We considered the alternative of using such sources in opportunistic fashion to produce, by electrolysis, hydrogen and oxygen at the offshore site, possibly combined with the extraction of other minerals from sea water. With the current interest in the use of hydrogen as a fuel for transport, there might be added value in this suggestion.

9.4.2 ACARD

ACARD (Advisory Council on Applied Research and Development) operated within the Cabinet Office and provided at least a feeling of being somewhere near the seat of power. The body was concerned for the most part with the preparation of reports on selected topics. At least in principle, and within the context of the variable political scene, the recommendations were understood to be reviewed internally to identify and anticipate possible objections by the Treasury or from elsewhere in Government. The remit of ACARD was essentially to advise Government on the effective application of 'applied research, design and development'. (The introduction of 'design' is interesting if undefined – it is probably intended to represent the intermediate processes between applied research and its commercial exploitation through development.) During my period of membership, I was appointed to chair a Working Group to prepare a report on the links between industry and higher education in the field of research and its application. In our report (ACARD 1983) we identified some of the barriers to co-operation, including historically, as may appear remarkable to present attitudes, reduced exchequer funding to universities who gained significant finance from other sources. We introduced the notion of a 'seed-corn fund' whereby, on the contrary, a reward of 25% of earnings through contracts, consultancies and research would be provided over a period of five years, with further support for the infrastructure for co-operation. We identified examples of success in co-operation. These included the Universities of Warwick and Cambridge.

Warwick is an interesting example of enterprise, particularly having regard to the fact that, a few years previously, it had been the centre of protest against any dealings with industry, fomented by the historian E. P. Thompson, author of *Warwick University Ltd* (Thompson 1970). We recognised what we titled the 'Cambridge Phenomenon', the extent of technological enterprises that had sprung up around, and associated with, Cambridge University. We recommended that a

Ethics and politics

study should be undertaken of this development, identifying the particularly favourable circumstances and how the lessons might be applied elsewhere. This study was undertaken and published (Segal Quince 1985), promoted and financed by a group led by Christopher Bullock of Barclays Bank, who also chaired the associated steering group.

9.4.3 *The launch of the International Tunnelling Association*

The most far-reaching example of my excursion into politics started from an invitation from the MoT to serve as British representative on a committee to be established by OECD (Organisation for Economic Co-operation and Development) to encourage a wider use of sub-surface usage, particularly by developing countries. I found that a number of old friends had been appointed to represent other OECD countries; we had mostly met through serving together on the Technical Committee for Tunnelling of PIARC. The preparatory programme provided an essential opportunity to appreciate the strengths and special interests that each brought to the meetings and to develop the mutual confidence vital for success of an international group. Once the objectives were considered, the need for action on an international scale seemed evident, and it was only curious that the initiative had had to come from an external body. On reflection, this is in fact a typical situation in which an external observer more readily identifies need for a culture change than occupational participants themselves. The representatives of the 15 nations who had accepted the invitation to participate met periodically at the OECD headquarters in Paris, planning several initiatives. We launched surveys on tunnelling techniques and on future predicted demand, making preparations for an Advisory Conference on Tunnelling in Washington in 1970 at which the state of demand and technologies were reviewed. Here, also, a series of resolutions were agreed, including the establishment of an international organisation concerned with tunnelling (OECD 1971). I had been appointed Chairman of the steering committee, so inherited the task of putting this next stage into effect. I place on record my admiration for the manner in which this preparatory stage was managed by OECD through their representative, Mr C. K. Orski. An interesting feature of our Washington meeting was that the word 'tunnelling' appeared frequently in our manifesto which I was drafting. This was before the days of word processors so I had the services of a group of typists who objected 'but mister you can't spell'. I needed continuously to insist on the second 'l' in 'tunnelling', which I claimed as my prerogative. So it is that the International Tunnelling Association

(ITA) (in French, the Association Internationale des Travaux en Souterrain (AITES)) – a slightly different connotation but interpreted in equivalent fashion – although conceived in Washington, is so spelled, and almost certainly why Japan has accepted the British form! The first need was to set up national 'focal agencies' towards international co-operation. The ICE undertook to provide secretariat and other resources for the purpose and a series of preparatory meetings followed, with much enthusiastic support from other countries. The British Tunnelling Society was established with Sir Harold Harding as first Chairman in 1971 within nine months of the Washington Conference. Sir Harold was largely responsible for the support provided by the Institution at this time, which certainly helped to enhance the position of Britain in international tunnelling. The inaugural meeting of the ITA was held in Oslo in April 1974, at which I was elected as first President. It was agreed by participants that Mr P. C. Beresford of the staff of the ICE should serve as interim Secretary. The nations represented were: Belgium, Denmark, Finland, France, Federal German Republic, Iceland, Italy, Japan, Netherlands, Norway, South Africa, Sweden, Switzerland, UK and USA. Initially, under my Presidency, statutes and bylaws were drafted in the UK and a French edition prepared by CETu (Centre d'Etudes des Tunnels) in Lyon. The Secretariat was established in 1975 at CETu. There was concern that there might be suspicion of our Association from countries outside OECD; we endeavoured to disarm such attitudes and we soon had members from behind the iron curtain and, later, China. We set out our statutes in a deliberately open manner, the acceptability for membership being based on an identifiable representative body for tunnelling within a country recognised as such by the United Nations.

Initially, a balance needed to be struck between bold initiatives to attract world-wide support and the raising of adequate funds through the membership. ITA is now well established as the international body for tunnelling, holding annual meetings by invitation in association with an international conference related to aspects of the use of the sub-surface. During the course of development ITA established close relationships with the American Underground Space Association and with the Society for Trenchless Technology (concerned with techniques for mini-tunnels for services). The ITA Journal, *Tunnelling and Underground Space Technology*, provides the vehicle for technical papers for the three organisations.

Much of the work of the ITA is conducted through Working Groups on themes selected by the membership. Each Working Group has a specific task, normally the preparation of a report, under the direction

Ethics and politics

of a Chairman and the guidance of a 'Tutor' who is a member of the Executive Committee of the ITA. Essentially, the objective is to encourage best practice in all aspects of underground planning, design, construction and maintenance. The operation of transport tunnels is recognised as the province of the international bodies for road and rail transport, with which a close liaison is maintained. In 1999 ITA celebrated its Silver Jubilee by a return visit to Oslo at which HM King Olaf gave a welcoming address.

There were occasional problems with protocol, in selecting between rival organisations wishing to represent one country and in ensuring that a country hosting the annual conference would provide visas for our members from, particularly, (pre-Mandela) South Africa. At the present time, there are 51 member nations and a number of individual affiliates. As Honorary Life President, I have observed and, on request, contributed to the development of the ITA. The degree of international co-operation has been outstanding. The main requirement at the present time, in my view, is a deeper integration between the work undertaken by the ITA and the corresponding activities in its member countries, particularly in the operations of the Working Groups. Understanding of the potential for going underground needs encouragement by an imaginative view of the opportunities. There is danger of excessive efforts being put into the 'business' of the organisation. There remains an urgent need for re-attachment to the fundamental purposes of the organisation, to prosper the art and appropriate application of underground solutions to economic, social and environmental problems. One of the interesting features of ITA is the manner in which the engineers, who form the dominant membership, have been joined by lawyers, concerned with the particular nature of planning regulations for underground space, and by architects who have developed schemes for optimising the multiple forms of using such spaces. Underground planning, with notable examples from Sweden and USA, was a central activity from the beginning. The physical benefits from going underground derive from even temperature, isolation from noise and vibration. The problems are those of access and in overcoming sensations of claustrophobia.

9.4.4 *Research Councils*
I have been involved in several activities of the research councils. I served on the Council of SERC (Science and Engineering Research Council), not long after the 'E' had been inserted, a period when the activities of Council were dominated by 'big science', particle physics

199

on the one hand and astronomy on the other. The common feature of the 'micro' and the 'macro' was that each relied upon a non-commercial client, the state, for its applications, at least at that time. What was equally apparent was that each was a major customer for highly specialised engineering of great complexity and precision. In fact, the high cost of 'big science' sprang essentially from its technology rather than its science.

To an engineer, it seemed odd that the Engineering of SERC comprised the technology but excluded management, the province of ESRC (Economic and Social Research Council). Having criticised this separation, I should not have been too surprised to be invited to serve as Chairman of the joint SERC/ESRC Committee, a body with a liaison role, no funds, little influence but much positive support from the participants. Apart from certain duties imposed on the Committee, we directed our main efforts to organising periodical workshops on themes which crossed interests of the parent Councils and which brought together people who would otherwise have been unaware of the complementary nature of each other's activities.

In 1981 I was invited by the Science Research Council (SRC as it then was) and the Departments of Environment and Transport to undertake, as Chairman of a Task Force, an assessment of long-term research needs and activities for construction. This report was published for limited circulation as 'Long-term research and development requirements in Civil Engineering'. Part of the exercise concerned the establishment of research priorities in different branches of civil engineering as perceived by practitioners. Another objective was to establish the value of research being undertaken for each type of activity. This was a complex task, which drew upon contributions from around 700 practitioners, specialists and researchers. It was not difficult to identify current research commissioned externally by an organisation. Much innovative activity takes place within the initiating organisation and this should be included in a normal definition of research. When construction is castigated for its low expenditure on research, this element, a most important feature, should be identified, even if difficult to evaluate in money terms. Another issue that was stressed is that there is a limit to the benefit from Government in construction research. A far better use of money stems from making innovative capability a feature in appointment of consultants and contractors, thereby enhancing the value put upon research – and leading to a wider benefit for the user public. In this way, Governmental research funding would be achieved predominantly by 'pull', confining

Ethics and politics

'push' largely to those areas of special interest to Government and remote from commercial application. We emphasised the need for public construction programmes to be planned as stable and essential parts of the economy and not 'as economic regulators to be turned off and on to meet short-term budgetary needs' (SRC 1981). We also addressed the increasing breadth of the civil engineer's activities in relation to social and environmental issues. We posed questions in these areas, leaving the answers to the long-term research we were recommending. The fact that this report was commissioned by the Government emphasised that the deficiencies in civil engineering research were at least recognised as a national problem.

9.5 The engineer, the politician and the public interest

The engineer, the politician, the public and the press are interrelated in a complex manner. It is often difficult to identify the common purpose of social improvement. Politicians are beset by the law of unintended consequences, despite the existence in the Cabinet Office of a group charged specifically to warn of such eventualities – but does this protection still exist or does a wilful Prime Minister, or his powerful advisors, take any notice of its warnings? The function is now presumably exercised through the Strategy Unit of the Cabinet Office (Strategy Unit 2002). The enthusiasm for the gratification of a short-term ambition, bringing expected popular support to its author, continues to obscure the recognition of the likelihood of the consequences conforming to the plan. Professional engineers, by their training and experience, bring objectivity to bear on foreseeing risk of this nature and might well provide material evidence, thus helping to avoid embarrassment to the politician (if he is still in place) and the attendant cost to the public. Government consults widely within the world of finance and commerce, also with the law, but these bodies provide poor advice in relation to risk management. Each is both risk prone and risk averse, preferring solutions which appear to transfer risk to others, whether appropriate or not, for which a high price is in due course levied on the taxpayer.

Where the policy for construction becomes distorted, with reliance placed excessively upon commercial practices and on least cost of each fragmented task, the engineer, private or public, consultant or contractor, is denied the opportunity to apply his expertise effectively. The engineer will be subsequently blamed for the resulting poor value for money by those who do not understand the causative processes of a dysfunctional system. Our immediate past history has erected a barrier to recovery of esteem for the ingenuity and objectivity of the engineer, to deserve the respect and trust that must attend his proper function in the public interest. The professional dimension needs to be emphasised, primarily in the interest of our Clients and the public.

In matters of general legislation, a reformed Upper House would contain leading figures from the professions, industry and the arts able to give advice freed from pressures from Party inquisitors. A function of the Press should be to provide publicity for these observations, which will not necessarily be seen by Government as politically correct. The engineering profession has collectively an ability to foresee events likely to follow from Governmental policy in the domains of industry, infrastructure and the environment. In advising Government, engineers should be satisfied to play a catalytic role to achieve results, not one of political, even if non-aligned, prominence, since politicians share the general antipathy to responding to correction by others.

The 'Manichean dilemma' for a social structure offers either a state with a central command structure or one which is wholly dependent on the market. The former is unacceptably inefficient, the latter the cause of a consumerist society polarised between the 'haves' and the 'have nots'. There is a subtle way in which the professions can assist in finding a 'middle way' for Britain. The US continues to provide a social model, albeit considerably tarnished, essentially based on free market principles, distorted by popularist measures to protect favoured groups in the home market. Measured by productive capacity and power, the objectives are achieved. The wide, and widening, gulf between rich and poor, with the associated violence and crime are features which most in Britain would wish to avoid. The issue of discretion in the use of hegemonic power is well beyond my theme. On the other hand, the central command economy used to have its power base in the USSR. There remains a considerable element of state control in the countries of central Europe. While undoubtedly the arrangement has ensured that much of the infrastructure is superior to our own standards, the degree of regulation contributes to bureaucratic costs. The middle way depends on finding a modification to the simple one-to-one relationships of the market place. It is here that the concept of the intelligent market, whose mechanism is described in Section 5.1, comes into play. The intelligent market depends essentially on professionals for its operation, as much in the medical and educational as in the industrial worlds. Professionals depend on

Ethics and politics

trust to establish value-for-money in place of the frustrated expectation of success from disintegrated projects on purely commercial relationships or to meet bureaucratically imposed targets. Egan correctly insists (Egan 1998) that the Client, whatever its form or constitution, must lead the project, corresponding to the Treasury notion of the enlightened purchaser. The question of trust lies at the heart, for example, of the professional engineer's concerns with risk. We are destined for economic decline if politicians were to share the view expressed by the Chancellor of the Exchequer that only those who work in the public sector are to be trusted not to sacrifice quality for profit.

At the present time, the National Audit Office pronounces with tedious regularity discoveries of gross waste of public money. Usually in retrospect, the cause is obvious, too often a result of failures in communication between Departments or failure to identify risk. Action, we are always assured, is being taken to prevent repetition, no heads roll, no Minister resigns, confidence in Parliament is a little further eroded. To an experienced professional trained in the area concerned – and the essence of a profession entails synthesis across the different issues – the risk of such loss will be evident in prospect, not only in retrospect. The politician should find the professional as an ally in such respects. A unified engineering body (Section 3.2) should be able to provide advice based on the best opinion across the engineering profession; this body should also be able to advise on the time-scales needed to introduce timely solutions to impending problems. Politics too often procrastinates to avoid awkward decisions, thereby imposing cost and discomfort on the public. A Chief Engineering Advisor in Government would be in a position to advise on the selection of critical issues meriting such external advice, which would be open for all to read.

The external impressions of the disintegrated nature of the engineering community, their separated and rival institutions, each immersed in solving domestic administrative and technical problems, gave in the past little encouragement to exploiting their knowledge in the wider problems of the state. More than 25 years ago, in my Presidential Address (Muir Wood 1978), I emphasised the broadening scope of the engineer's functions in social and environmental affairs, including the notion of sustainability. These developments have now taken place. With the rising impetus towards unification of the engineering profession and the several initiatives taken by the unifying bodies that now exist, there are mechanisms that can readily be tapped to provide advice of benefit to politicians in framing legislation. Current initiatives by the Royal Academy of Engineering relating to risk and to expediting the application of lessons to be learned from accidents – or near misses – represent the type of engineering know-how of benefit to the politician.

9.5.1 Private and public

The principle of private funding for facilities providing a service for the public is not new. The roads (turnpikes), canals, gas and electricity supply, and the railways in Britain were all financed, with more or less success, by such means. So, often through Trusts operated mainly by the immediate users, were dock and harbour authorities. Water supply was organised both by city authorities and by private companies. Health and education became partially public services, through private origins. The first fire services were run by insurance companies. The entrepreneur sees a market, promising a return in excess of investment, and plans to fill it. As people come to rely upon the service, so are conditions attached to assure, or at least encourage, security of supply.

A tenet of political faith for the control of the 'commanding heights' of industrial society led on, in the 1940s, to the nationalisation of transport, power, water, and the steel industry with motives little clearer than those of political control. Certainly, there was no discussion in relation to gains in efficiency.

At the present day, privatisation is favoured by those who perceive nationalised industries: as operated for the benefit of its employees (whose numbers are seen as excessive) in preference to the user; and as lacking incentives for efficiency and innovation, coupled with inability to benefit from enterprising leadership. To these are added the more nebulous notion of reducing charges against the Government's direct borrowing.

As already discussed in Chapter 5, the concept of partnering, demonstrating benefits in good engineering, purposeful continuity and innovation, all contributing to much improved value for money, broadens the professional element of each engineer's function. Generally, PFI – used here in a shorthand sense to include all forms of transfer of financial responsibility for public projects to the private sector with a formula for payment over a period of years – has yet to be established on a sound conceptual basis. It is first necessary to consider its objectives and the causes of excessive cost, failings in quality and loss of control of cost. For this simple analysis, the financial

Ethics and politics

jargon of special purpose vehicles, mezzanine finance and the several forms of equity investment, as the financial means for achieving objectives, may be set to one side. The stated objectives are:

- To enable projects to be undertaken which would otherwise cause excessive demands on the taxpayer or on Government borrowing.
- To exploit superior ability of the private sector in the management of such projects.

These are, in fact, wholly separate issues, which will in consequence be treated as such. PFI places an increased cost (25%?) on the taxpayers of future years; in equity between the generations, therefore, PFI should be confined to projects conferring long-term future benefit. In logic, also, PFI is only an option up to a limited aggregate in magnitude and time; otherwise it contributes to an excessive future national debt in relation to gross domestic product (GDP). Currently (2003) the 'off account' volume of PFI projects, spread and not co-ordinated among different Departments, is estimated to amount to about £100 billion. Only within limits, therefore, is PFI sustainable, and the principle acceptable in equity. The several options for borrowing funds do not in themselves predicate PFI. Problems that remain to be solved are many. The most troublesome concerns the achievement of the objective of maximising satisfaction to the user, for the output of an organisation, providing a project wholly defined at the outset, whose primary stated objective is to satisfy the shareholders. In the place of the *raison d'être* of a statutory authority arise new controlling bodies and regulators, in effect attempting to inject vicariously Adam Smith's notion of the 'Invisible Hand' to control the wilder excesses of the free market.

The benefit of private sector management deserves far closer scrutiny. In principle, the devolution of matched powers and responsibilities to those who have the skills to apply them must be a positive feature. The 'boundary conditions' of the process are the features that demand particular exploration. We first need to consider the issue of risk. PFI involves a transfer of responsibility for project control at a particular moment in time. Part of the problem that leads to excessive cost stems from premature transfer; as a consequence, the extent to which uncertainties may subsequently develop as hazards with inestimable incidence of costs has to be accepted as an inherent cost of PFI.

An alternative strategy would be far preferable in the public interest. For each project, the process of risk assessment and management needs to be established prior to transfer, the register of potential risks allowing action in their investigation, elimination or reduction to be undertaken.

There is no public interest in paying for risk which is beyond the control of the project team and which may not eventuate; this simply adds unnecessarily, and often considerably, to cost. It is particularly expensive and ridiculous to expect the PFI undertaker to carry elements of risk which are under the control of the Client or any wing of Government. At the moment of transfer, the PFI undertaker has to take a pessimistic view, a feature which adds greatly to the cost where uncertainty is high.

The second feature that could benefit from advice from professional engineers concerns the effect of variation, another major cause of increase of cost of PFI projects, where this has not been foreseen and provided for. Variation in this context may be seen as a form of risk but not one transferable to the private performer. Variations can cause major disruption to the contract programme and sequence of operations. The nature of possible variation is foreseeable by those with experience with comparable projects. The possibility may be included in the provisions of a PFI agreement, including, for example, the latest acceptable date for a specific change in relation to a construction or other programme. Renegotiation of a PFI project which runs into unforeseen problems of such nature is an expensive business for a public Client in a particularly weak negotiating position.

Some PFI projects place excessive responsibility upon the provider for issues that depend intrinsically on quality that cannot be fully specified. At the Royal Military College of Science at Shrivenham, for example, a large new block is not only being built under PFI, but the provision of teaching is also included, as is the risk of continuing use when the period of the PFI repayment ends. These features seem likely to cause problems which would have been avoided by confining the PFI element to the bricks and mortar.

9.5.2 Railways

The railways provide an outstanding example of mismanagement by Government, for which a high price remains yet to be fully paid before standards can be expected to match those expected as acceptable by comparable European countries. Our railway history depended on private railway companies associated in a series of amalgamations prior to the Second World War. Nationalisation followed in 1946, after a high degree of Government support to keep the railways functioning through the war. The Beeching axe of the 1960s led to reductions in the length of track.

Ethics and politics

Public infrastructure demands a 'consistency of purpose not often seen from Britain's Governments' (Report of the Strategic Rail Authority 13.3.01). Prior to privatisation in 1993, it was well known to railwaymen and others that the railways had been starved of the funding necessary to recover the backlog in essential maintenance and modernisation of an ageing system. Political parties did not address the fundamental factors for the success of a railway, however owned or funded. The party in power wished to achieve the transfer predominantly for political reasons of minimum public control. The opposition failed to mount an effective attack on the obvious technical failings of the fragmentation of the proposal, unworkable for an objective for massive capital injection into different, but complementary, features in a variety of private companies. The empty threat of renationalisation only added further loss to the taxpayer by reducing the value of the assets as perceived by the purchasers. One of the first consequences was the resale of several of the franchises, providing high immediate gains for those buyers who had been prepared to dismiss renationalisation as a political gimmick. Another consequence was the dismissal of large numbers of experienced railwaymen, whose combined experience is, at the least, costly to recover.

After several accidents attributable at least in part to organisational failures, steps have been instituted to recover the situation, the responsibility for the infrastructure being transferred from Railtrack to Network Rail. Plans for the finance of major schemes through PFI (here termed DBFI – design, build, finance and transfer), as well as massive increase in funding by Government, are proposed. As a result of the continuing parliamentary incompetence identified by Robert Stephenson in relation to the railways (see Chapter 1), the vision of an adequate railway system yet remains a distant prospect. Meanwhile, subjection of the fragmented organisations for infrastructure, operation and train sets to external management by the Strategic Rail Authority (SRA) and the Railway Regulator, with different objectives, ensures that there is an excessive bureaucracy obstructing the functions of management in attaining the optimal balance between safety, efficiency and reliability. Costs are increased greatly by this superstructure, with increasing public scepticism as to whether increasing subsidy is justified by the rate of improvement. The organisation remains a long way removed from the concept of the railway perceived and operated as a system, in the control of experienced professionals who set the balance between cost, service and risk.

Civil engineering in context

9.5.3 Energy

There are few areas of such concern for the concept of sustainability than that of energy. The politician, advised by economists, pursues policies dominated by GDP and rates of growth, as if these represented symbols of virility. The engineering profession must follow two principles:

1. To develop clear advice on practicable energy planning for the future, to establish 'degrees of freedom' for politicians in formulating energy policies.
2. To campaign for pre-market research and development to be commissioned, as may be needed to implement energy policies economically and efficiently.

So long as ACORD survived (Section 9.4.1) engineers might reasonably expect this body to be providing sound advice in both these respects.

In my 1982 Unwin Lecture (Muir Wood 1982), I provided a current view of the capital and running costs of the several sources of primary energy for power generation. I stressed the special features of renewable sources in terms of high capital cost and their specific degrees of reliability, e.g. dependence on the weather. By a simple application of 'energy accounting', I also provided an approach to the maximum permissible exponential growth rate of a renewable source if it is to contribute to savings from other sources of energy.

Professor B. J. Brinkworth has drawn attention to the historical evidence of the relationship between energy intensity (i.e. demand per capita GDP) and time, indicating that during intermediate stages of development this ratio reaches a peak and subsequently declines. As a world strategy in energy planning, one object must be to reduce this 'hump'. The effect is illustrated by Fig. 9.1, derived from Martin (1988). It will be noted that the later in time that industrial development has occurred, the lower the 'hump'.

At a seminar in 1994, organised by the Watt Committee of the Engineering Institutions, I emphasised, in my task of summarising the occasion, the need for a 'Government strategy to identify where research and development need to be conducted to ensure that the promising options are kept open for timely exploitation – on a timescale, however, which would not attract market-driven support'. This was dangerous heresy, however, at a time when the free market was expected to solve all the energy problems – which it did in the short-term by maximising the use of natural gas. A strong objection was

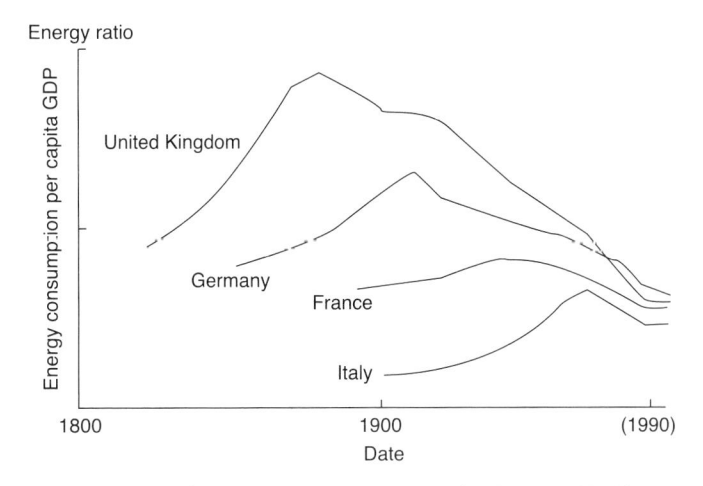

Fig. 9.1 Development of energy ratio in Europe (After Martin 1988)

lodged to my observations by a representative of the DTI. There was a delay in publication, by which time Government had changed and my remarks were then seen to be nearer the canon (Watt Committee 1999).

Government has now accepted responsibility for a strategy, but has yet to give evidence of heeding advice from the engineering community. This advice stresses the need for diversity, the essential element of nuclear power and the practical limits of wind energy, the currently favoured renewable. Any intermittent source requires surplus generating capacity in modes without such a handicap, an investment which needs to be included in any economic assessment. Wind energy, notwithstanding its appeal for apparently gaining something for nothing, requires in consequence a high subsidy to compete with other modes. Can Government be trusted to maintain these subsidies indefinitely? At a time of suspected changes in weather patterns, possibly related to climate change, particular care is needed in computing the risk, taking account of the geographical distribution of the wind-powered generators. The current plan by Government would depend to a very high proportion (virtually completely when the wind does not blow) on natural gas, mostly imported from afar. This seems to be a plan based on hope – including charitable treatment by suppliers once the degree of dependence on their supplies is established – rather than discernment. One day, politicians will need to overcome their excessive deference to the myopic anti-nuclear lobby, which does not need to accept responsibility for future power supplies. New, or refurbished, nuclear power stations need to be sponsored before the existing stations, and those with appropriate skills, are retired.

9.6 The environment and sustainability

Engineers accept Environmental Impact Assessment as a fact of life, albeit a highly bureaucratised means of protecting the environment. The technique derived from the US as a result of the flagrant disregard for the secondary effects of industrial processes extending into the second half of the twentieth century. This was as serious as the effects in the nineteenth century in Britain, when ignorance and attitudes provided better excuses. The Sierra Club is a respected body for environmental protection in the US. Even this body can be derailed by the free market. Having lobbied successfully against additional hydro-power plants on the Colorado River, a power deficit became apparent. This, therefore, required urgent construction of new coal-fired plants. Only later was it known that mining interests had helped to finance the anti-hydro campaign. This is a constant danger for single-issue bodies with good intentions but lacking a rounded view. Engineers who balance micro-sustainability issues for individual projects are constantly assailed by the macro-unsustainability sponsored by politicians by conflict, encouragement of boundless subsidised air travel, with its boundless release of effluvia and many other thoughtless examples of profligate and conspicuous waste, many arising from the 'law of unintended consequences'.

In his 22nd Graham Clark Lecture (Ashby 1978), Lord Ashby recommends attention to historical evidence as to how people adjust to the secondary consequences of major engineering achievement. He describes the initial stage of increasing perception of environmental hazard or nuisance, leading to attention to the practicalities of abatement without unacceptable industrial damage. The result was the phrase 'best practicable means' (bpm) whose significance changed with time from a protective arrangement for the polluter to a criterion for conformance, by those applying standards. This step was associated with the Alkali Act of 1863 which set standards measurable by scientists, who could interpret bpm, thus avoiding the heavy-handed imposition of the law. The Inspector was allowed discretion, while control limits were devised for the pollutants. In 1956, the Clean Air Act added the provision 'having regard, among other things, to local conditions and circumstances, to the financial implications and to the

Ethics and politics

current state of technical knowledge'. These provisions allowed the Inspector to permit major improvements in control to be undertaken for new construction in conjunction with other aspects of modernisation. Smoke control was an especially difficult case on account of absence of precise definition, until the Clean Air Act established smoke control zones with the application of the Act through Local Authorities. Here, Lord Ashby commends the conjunction of bpm with local controls through the intermediary of scientists and engineers. So long as politicians confine themselves to general issues of policy, all well and good. If, however, these principles are translated into codified statements of centralised application, regardless of the differing circumstances, the ratio of cost to benefit will be unjustifiably high. This seems to be an instance in which the pragmatical approach of Britain is superior to the 'Cartesian elegance of a harmonised environmental policy for Europe' (Ashby 1978). There are several morals to this story, for more general application:

- Progress in environmental improvement may be undertaken by balanced foresight; this may be more effective than militant demonstrations.
- Local discretion is beneficial in determining appropriate balance between an economically desirable development and its negative consequences.
- General legislation should provide opportunity for decisions to be based on judgements of qualified professionals, not as dogmatic application of rigid standards.
- The public can be helped in forming balanced views on proposals affecting industry and the environment, through improved education on the science behind such decisions, also by greater efforts in the community to explain and demonstrate, through exhibits and the media, the breadth of practicable options. Improved degree of information would much reduce the impact of 'single issue' militants who do not need to address the practical consequences of acceptance of their opposition.

It should be obvious that environmental impact is an area in which legislation derived from the EU may have justifiably different degrees of applicability for Central Europe than the North of Scotland. For example, control of pollution of the Rhine requires a unified approach for all riparian countries. The achievements of the Clean Air Act and its predecessor, the alkali act, need to be better understood by politicians, lawyers and the 'greens' who are all given towards excessive codification. However controls are administered, the operation needs to be undertaken transparently, so the 'trade-offs' may be understood.

One specific feature of the approach described above is a differentiation between standards expected immediately for new construction and those for existing facilities. For example, under legislation to take effect in 2005, public buildings must be accessible to the incapacitated. Failure to comply could result in prosecution. For a village hall in precarious financial circumstances, serving as a meeting point for an isolated community, closure may be the only option, whether or not there are any disabled villagers. Far better to have applied bpm and to await the raising of funds for a major refurbishment, meanwhile relying on assistance from neighbours.

9.7 Broader reflections

The notion of sustainability is a popular rallying cry for politicians. Sustainability is another holistic concept, including safeguards for nature, society and prosperity. It is clear that we need to address our problems in a joined-up manner:

- Ensure that education, training, reward and opportunity are such as to attract recruits into the vital professions and skills.
- Ensure that training and retraining facilities are available for those who possess potential skills which are currently untapped.
- Encourage, by addressing local problems, industry, commerce, and all the features that contribute to a thriving community, into areas currently of deprivation, at the same time reducing pressures on the socially overloaded parts of the country.

The engineer needs to demonstrate to the politician the degree of shared interest in engaging in actions which have been exposed to reliable engineering advice on their practical soundness. The politician needs to respect the professional basis of the engineer in setting objectivity ahead of self-interest. While leading engineers have the responsibility to demonstrate that they live by their own maxims, the Institutions, led by the Engineering and Technology Board (or unified policy representing the engineering institutions), should drive such a campaign, by example as well as precept.

A major need is for the better education of the general public, so that they are less gullible to the sillier and more destructive campaigns of the press. We here return to the concept of trust, whose absence is so

Ethics and politics

destructive for those who try to work in the public interest. Trust depends on education so that the general public have a better understanding of science, can follow an argument based on simple mathematics or logic, who learn to discard the notions of anti-science based on ignorance and prejudice. I argue throughout this book for the benefit to be derived from a greater transparency attending the work of the professions. We will not overnight remove the deep-dyed cynicism of the press; we can hope to render it, by improved understanding by the public of the potential of science and engineering, less able to sway opinion into damaging cul-de-sacs.

There is a deeper malaise in that politicians are generally ill-equipped to make policy decisions dependent on an understanding of science and engineering. There is evidence that, however well prepared may be the brief on which such decisions need to be based, the actual moment of decision is rushed and conditioned by the political requirement of least offence to an ill-informed public. The common requirement from politician and engineer is to educate and inform the public, who would learn at the same time about rational concerns, thus better placed to understand the practical constraints under which political decisions affecting their well-being need to be taken (Bray 2004).

It is undeniable that the performance of politicians of all parties is held by the general public to fall well below promises, and even expectations. This is no place for an attempt at a fuller analysis but it is possible to identify, as seen from the viewpoint of a practising engineer, a few of the main contributory factors.

The British disease throughout the public and private sectors is the failure to subscribe adequate capital in terms of investment and training. For the private sector, financial transactions have provided a readier source of easy winnings than the more difficult and certainly more socially desirable functions of productive investment, which requires greater patience to await returns. The share-holder is generally fickle, expecting short-term profit, regardless of the absence of strategy for the longer term. It could be argued that a 'damping' mechanism, inserted between industry and the punter could encourage a longer view. In Germany, for instance, this intermediary has been provided by the industrial banks. Added to this pressure for short-term results is the spectacle of extravagant rewards, earned as readily by failure and premature dismissal as for success. Once a society observes significant lack of correlation between effort and reward, social cohesion gives way to cynicism. It is not so much the undeservedly rich pickings in themselves but that these siphon away the funds for potential productive investment. What happened to the railways in 1993? Their resources were scattered, immediate fortunes made. A disaster for the 150 years of early enterprise and later neglect.

One of the heroes to the orthodox investment world of post the Second World War Britain was Lord Weinstock, credited with the accumulation of riches in combining several electrical engineering companies into GEC (General Electric Company). What is forgotten is that along the way, this activity effectively destroyed Britain's capacity for heavy electrical engineering. At least he conserved his capital, even if it was inadequately invested in productive industry. His successors leapt onto the electronic band-wagon with extraordinary ill-timing, its losses about to drain the irrational expectations. In this way, much of our electrical industry, and the careers of those who worked in it, evaporated into financial manipulation. This is a part of the general story of industrial decline.

For Government, the prime objective appears to be the economy, held out as the principal 'good' without concern as to how this may correlate with personal well-being. In principle, the health of the economy is measured by GDP or similar quotient. When we find that matters affecting quality are constantly sacrificed on the altar of quantity, we are entitled to stop to question the argument, not to buy off our own inadequacies by such short-term expediency 'for the economy'.

We can no longer afford policies which deprive a high fraction of the population of hope, expectations or skill. The bulging prison population of over 70 000, many illiterate and few with skills to earn an honest living, is a symptom, representing only a token of a wider malaise. Until we as a nation have seriously addressed the basic problems of satisfactions in life, education and work, we should grossly scale down our enthusiasm for military adventures overseas.

Suppose we were to recognise that the heroes of the day are those who invest in creating work, encourage a fresh appreciation of life through the acquisition of skills, and generally enhance perceptions of quality, we could create a society whose rewards followed such activities. There is indeed a middle way between the free market and the centralised command economy. The present compromise is a muddle. Increasing powers are devolved on commercial terms to private control of infrastructure, but scarcely enterprise, in view of the accompanying restrictions. Apart from much confusion in relation to risk, wasteful effort is then diverted to tiers of target setters, inspectors and external control, with corresponding efforts to achieve performance targets which affect reward, while ignoring wider achievement of

Ethics and politics

desirable goals. If we look back at Fig. 5.4 we can identify some of the factors that distort the three tiers of conversion from policy to execution into a far more complicated structure, with low value the outcome. Policies are not defined in association with those charged to plan their achievement and competent to do so, nor is the responsibility passed on in a clear hierarchical manner. So long as the structure depends solely on commercial factors it will not work. The overriding motivation will be short-term profit, not the achievement of political objectives. By inserting the professional element of motivation, policies are transferred into actions by those who accept responsibility for the results, balancing the several economic and non-economic factors, not managed like puppets on strings by external administrators and regulators.

It has been said that economics has no law like the conservation of energy to prevent the creation of purchasing power. Adams and Duckworth (2002) point to the collision course between market forces and sustainability. John Stuart Mill had foreseen this potential problem in 1848 (Mill 1848 bk IV, Ch VI sec 2) thus; 'It is scarcely necessary to remark that a stationary condition of capital and population implies no stationary state of human improvement. There would be as much scope as ever for all kinds of mental culture, and moral and social progress; as much room for improving the Art of Living, and much more likelihood of its being improved, when minds ceased to be engrossed by the art of getting on. Even the industrial arts might be as earnestly and as successfully cultivated, with this sole difference, that instead of serving no purpose but the increase of wealth, industrial improvements would produce their legitimate effect, that of abridging labour.' A subsistence economy, as have been many rural economies over the centuries, maintained an equilibrium with little change and reasonable contentment. The Free Market depends upon making the potential customer dissatisfied, appealing more to the emotions than to the mind to overcome this inadequacy by increased spending, into the realm of conspicuous consumption. Where basic needs are met, the next level of Maslow's pyramid of satisfactions (Maslow 1970) concerns comfort and well-being; beyond this level, satisfaction is increasingly related to education, both in performance and in the appreciation of performance by others – in art, literature and sport, for example. As engineers, we cannot support an increasingly self-indulgent throw-away society, justified by its contribution to ever increasing industrial output. This would imply an accelerating instability, consuming increasingly large quantities of wasting assets, encouraged by advertising and Government to improve GDP. In a rational world, with demands for action to prevent further degradation

Civil engineering in context

ever before our eyes, the promotion of unwanted consumption is insupportable. Increasing GDP is justified by raising the levels of the poorest and most deprived, here and overseas, also in conjunction with an element of redistribution, in rewarding those who are obviously underpaid at the present time. Part of the problem then becomes the fact that expected standards of living are only affordable for some as a result of the fact that the majority cannot afford to compete. In an era of peace between civilised nations and larger groupings to police the uncivilised, the desire for an increasing population seems to be nonsensical. Who actually wants (apart from the Chancellor of the Exchequer) a rising GDP and increasing population unless these contribute to the health and happiness of mankind, now and in the future?

To an engineer, the combination of encouragement of increasing immigration and employment in areas of failing infrastructure is a clear example of absence of system. At the present time, immigrants with professions and trade skills are being encouraged into areas where available housing and infrastructure deter UK residents. This is self-defeating and unsustainable. Those who have devoted their careers to training scarce professionals in developing countries are appalled – as are many others – to find that we now appear, as a matter of policy, to be attracting these same professionals to migrate to Britain to compensate for our lack in making adequate provision. This is grossly unethical. In any event, either these people will remain unassimilated or they will acquire our own expectations and, within a generation, add to the problem rather than the solution.

Perhaps the most extraordinary, economist prompted, policy is that of deliberately encouraging further inwards migration for demographic reasons, i.e. to provide more young people to support increasing numbers of pensioners. This may, at least in part, represent political rationalisation of the fact that control of immigration has been so lax that we do not know the numbers of illegal immigrants, nor even the identities of those who have overstayed temporary visas. A moment's reflection indicates that once such a solution is accepted, it must be continued as an exponential rate of immigration to a positive exponent. This implies that today's excessive rate of immigration, excessive in the question of ability for assimilation, must need to be increased – indefinitely! The social consequences are dire. Far better would be the provision of enjoyable employment for those who remain active well beyond the age of formal retirement.

Of course, we must accept our share of political refugees and, against absurd opposition, ensure that they are taught to use our language, be offered training and join our society, keeping their customs but not their isolation. In discussions on ethics with educated people of different religions as I have travelled around the world, I have found a remarkable degree of correspondence. For a lecture on energy planning in Bangladesh I deliberately inserted a quotation from Rabindranath Tagore, knowing him to have been equally revered by Hindus, Moslems and Christians.

If 'religious' schools concentrated on teaching ethics as a way of life in preference to fundamentalist diatribe and the teaching of stories of doubtful provenance and unexplained probity, a main feature of division between communities could be removed by shared education. Otherwise these minority groups cannot take an effective part in society, will be exploited and become a cost rather than an asset.

Having sold off much of the stock of council houses, we now find in certain regions of the country absence of housing affordable by those on low incomes, on whom vital services depend. The political reflex is to build large numbers of new houses in these areas, of which a fraction would be 'affordable'. There appears to be no attention to combating the forces attracting ever more residents to the most overcrowded areas, coupled with inducements to find work and better living conditions elsewhere. Present centres of growth cannot compete indefinitely against blight, overwhelmed services and general degradation that accompanies overcrowding. There are already signs that congestion around London, including the system of charging traffic entering the city, is encouraging organisations to move away. A lead from Government offices in the same direction could well be a highly constructive demographic action.

Another remarkable feature of the present is that slack regulation on loans (for the benefit of the economy!) has resulted in a level of personal debt increasing at an unsustainable level, currently (2003) about 12% of GDP per annum, several times the rate of economic growth. This accompanies rising national debt, also debt for posterity through PFI, of another 8% or so of GDP. The element of personal debt is the most potentially explosive, under persuasion to spend money people do not have on items they do not want. Will it lead to apprehensions triggering decline in property value, 'negative equity' and problems for the lenders, or will it provoke a brake in the form of increased bank rate, with a crisis as attempts are made to reduce borrowing? Either way, it provides yet another example of the inadequacy of the free market in unequal transactions between buyers and sellers. Why those who lend money on the security of inflated property values do not share the responsibility for the consequences is an unanswered question. The unsustainable increase in the level of personal debt is encouraged by Government to establish a rate of growth (in GDP) marginally ahead of our European neighbours!

These are important questions. Engineers are concerned with the manifest inefficiencies of our society; our familiarity with systems could help towards major reforms. The essence is that our professional competence and integrity be trusted.

10
CODA

This brief conclusion provides opportunity to restate and recombine my principal themes. In order to plan towards the goals I set for the civil engineering profession, we must understand our point of departure. Chapters 1 and 2, with allusions elsewhere, set out to provide this background.

Professions, at least in the British tradition and model, depend on the soundness of their institutions and their ability to adapt to change. Chapter 3 identifies some of the principal weaknesses and proposes reforms. Chapter 6 extends this analysis to include the 'formation', i.e. the education and training, of the civil engineer.

Chapter 4, and this is a recurrent theme, concerns the relationships between the numerous, currently too numerous, 'stakeholders' to a Project: the parties, who may be engaged through contracts or otherwise: representatives of interests internal or external to the Project; internal and external agencies; those affected by the Project and their representatives. Lawyers have inflicted much damage to relationships and inflicted much unnecessary cost, by assuming roles, beyond those for which they are trained, that they are incompetent to fulfil. The most serious problems with the law arise through their liability to create risk and their obstacles to its management.

Chapter 5 discusses the heart of the engineer's function, that of managing systems. Capabilities in this respect allow the civil engineer to penetrate far beyond his traditional roles of providing and maintaining infrastructure. After mostly reminiscence of Chapters 7 and 8, Chapter 9 relates the professional role of the engineer to ethics and applications to policy. The claim is made that only by full encouragement and exploitation of professional commitment (and this has general application to the productive professions) can we avoid a society which depends either on the inhuman inequalities of the free market or the petrifying grasp of the centrally planned economy. At present we suffer from many defects of each.

One central theme appears under Section 9.6. Currently politicians, under legal advice, convert desirable reforms into rigid standards supported by legislation, leading to great waste of time and resource in the argument arising from manifestly absurd prosecutions. This has also entailed a pathological aversion to risk which threatens livelihoods founded on common sense and even the survival of industries. Far better, is the advice, by Lord Ashby 25 years ago, with general application, to enact general legislations, but leaving decisions, to be made in the open, by qualified professionals.

Throughout the book, as a ground basis, runs the underlying notion of trust, which needs to be continuously earned and retained by our profession through the demonstration of high standards of wise, principled and creative leadership.

References

ACARD (1983) *Improving Research Links between Higher Education and Industry* (Advisory Council for Applied Research and Development in collaboration with the Advisory Board for the Research Councils). HMSO, London.

Adams, K. G. and Duckworth, W. E. (2002) Towards sustainable growth, *Ingenia*, **11**, 60–2. Roy. Acad. of Eng.

Armstrong, J. H., Dixon, J. R. and Robinson, S. (1999) *The Decision Makers: Ethics for Engineers*. Thomas Telford, London.

Arup. (1992) *Education for the Built Environment*. Arup, London.

Ashby, E. (1978) *Engineers and Politics: A Case History*, 22nd Graham Clark Lecture, 26 January, Council of Engineering Institutions, London.

Baker, J. F. (1951) Developments and trends in university education in engineering, *Joint Engineering Conference (ICE, IMechE, IEE)*, 261–4.

Banwell, H. (1964) *Report of the Committee on the Planning and Management of Contracts for Building and Civil Engineering Work* (Chairman: Sir Harold Banwell), HMSO, London.

Barnes, M. (2000) Civil engineering management in the Industrial Revolution, *Proc. Inst. Civ. Engrs*, **138**, 135–44.

BCSA (1986) *The Theory and Practical Design of Bunkers*, British Constructional Steelwork Association Publication No. 32, London.

Blockley, D. and Godfrey, P. S. (2000) *Doing it Differently*, Thomas Telford, London.

Bray, J. (2004) *Standing on the Shoulders of Giants*, E. Bray, Cambridge.

Bridges, G. P. (1951) Discussion on design and construction of silos and bunkers, *Inst. Civ. Engrs Struc. & Bldg Eng. Paper 27*.

Bruckshaw, J. M., Goguel, J., Harding, H. J. B. and Malcor, R. (1961) The work of the Channel Tunnel Study Group 1958–60, *Proc. Inst. Civ. Engrs*, **18**, 49–178.

Buchanan, R. A. (1989) *The Engineers. History of the Engineering Profession in Britain 1750–1914*. Jessica Kingsley, London.

Chrimes, M. M. (2004) British civil engineering biography, Part 1: 1500–1790. *Proc. of Inst. Civ. Engrs, Civ. Engng,* **157**, 91–96 (May).

CIRIA (1978) *Tunnelling – Improved Contract Practices.* Report 79, Construction Industry Research and Information Association, London.

Cockcroft, W. H. (1982) *Mathematics Counts,* HMSO, London.

Egan, J. (1998) *Rethinking Construction,* HMSO, London.

Ewing, J. A. (1926) *The Steam Engine and Other Heat Engines,* Cambridge University Press, Cambridge.

FIDIC (1987) *Conditions of Contract for Works of Civil Engineering Construction,* 4th edn, Fédération Internationale des Ingénieurs Conseils, Lausanne.

FIDIC (1999) *Suite of Contract Documents,* Fédération Internationale des Ingénieurs Conseils, Lausanne.

Finch, J. K. (1951) *Engineering and Western Civilisation,* McGraw-Hill, New York.

Finniston, H. M. (1978) *Report of Government Commission of Enquiry into the Engineering Profession,* HMSO, London.

Flanagan, R. and Norman, G. (1993) *Risk Management and Construction,* Blackwell Science, London.

Florman, S. C. (1976) *The Existential Pleasures of Engineering,* St Martin's Press, New York.

Harper, B. C. S. (1996) *Craft to Applied Science. The Institution of Civil Engineers, London and the Development of Scientific Civil Engineering in Britain, 1818–1880,* Ph.D. Thesis, University of Melbourne (August).

Harris, C. S., Hart, M. B., Varley, P. M. and Warren, C. D. (eds) (1996) *Engineering Geology of the Channel Tunnel,* Thomas Telford, London.

Highways Agency (2001) *Highway Agency Procurement Strategy,* Highways Agency (November).

Hudson, K. (1963) *Industrial Archaeology,* Waterlow, London.

Humphrey, J. W., Oleson, J. P. and Sherwood, A. N. (1998) *Greek and Roman Technology,* Routledge, London.

ICE Archives Panel (2000) *Biographical Dictionary of Civil Engineers of Great Britain and Ireland 1500–1830 (BDCE).* Thomas Telford, London.

ICE (1993) *The New Engineering Contract,* Thomas Telford, London.

ICE (1996a) *Sustainability and Acceptability in Infrastructure Development,* Thomas Telford, London.

ICE (1996b) *Whither Civil Engineering,* Thomas Telford, London.

ICE (2000a) *Royal Charter, Bylaws, Regulations and Rules,* Inst. Civ. Engrs, London.

ICE (2000b) *Annual Review,* Inst. Civ. Engrs, London.

Inglis, C. E. (1941) Presidential address, *Jnl Inst. Civ. Engrs,* **15**, 1–18.

Kidd, B. C. (1976) Instrumentation of underground civil structures, *Proc. Symp. on Exploration for Rock Engineering,* Johannesburg, 210–31.

Landels, J. G. (1978) *Engineering in the Ancient World,* Chatto & Windus, London.

Latham, M. (1994) *Constructing the Team: Final Report of the Government/Industry Review of Procurement and Contractual Arrangements in the UK Construction Industry*, HMSO, London.

Legget, R. F. (1982) *John By*, Historical Society of Ottawa, New Brunswick.

Legget, R. F. (1984) *Engineering in Canada: Past and Present*, University of New Brunswick.

Lemoine, B. (1991) *Le Tunnel sous la Manche*, Editions du Moniteur, Paris.

Lightfoot, E. and Michael, D. (1967). Prismatic coal bunkers in structural steelwork, *The Structural Engineer*, **44**, 55–62.

Maslow, A. H. (1970) A theory of human motivation, in *Management and Motivation* by V. H. Vroom and E. L. Deci (eds), Penguin, Harmondsworth, 27–41.

Mainstone, R. J. (1998) *Developments in Structural Form*, 2nd Edn, Architectural Press, London.

Martin, J. M. (1988) L'Intensité Energetique de l'Activité Economique dans les Pays Industrialisés: les Evolutions de Très Longue Periode Livrent – elles des Enseignements Utiles? *Economies et Societés* – Cahiers de L'ISMEA, **22**(4), 9–27.

Melchior, P. (1966) *Earth Tides*, Pergamon.

Mills, J. S. (1848). *Principles of Political Economy*. Parker & Co, London. (Many subsequent eds).

Morgan, H. D., Haswell, C. K. and Pirie, E. S. (1965) The Clyde Tunnel, *Proc. Inst. Civ. Engrs*, **30**, 291–322.

Morley, J. (1994) Importance of being historical: civil engineers and their history. *Jnl Prof. Issues in Eng. Educ. & Prac. (ASCE)*, **120**, 419–28.

Muir Wood, A. M. (1955) Folkestone Warren landslips: Investigations 1948–50, *Proc. Inst. Civ. Engrs Pt. 2 (Rly. Paper 56)*, 410–28.

Muir Wood, A. M. (1970) Characteristics of shingle beaches, 11th Coastal Eng Conf., Washington, 1056–76.

Muir Wood, A. M. (1978) Presidential Address, *Proc. Inst. Civ. Engrs*, **64**(1), 1–23.

Muir Wood, A. M. (1982) Energy research – whence and whitherwards, *Proc. Inst. Civ. Engrs Pt. 1*, **72**, 285–305.

Muir Wood, A. M. (1991) The Channel Tunnel – view of a teredo, *Tlling & Und. Sp. Tech.*, **6**(1), 77–82.

Muir Wood, A. M. (1994) Geology and geometry: period return to Folkestone Warren, *XIII ICSMFE Conf., New Delhi*, Vol. 5, 23–30.

Muir Wood, A. M. (1995) *The First Road Tunnel*, Permanent International Association of Road Congresses, Paris.

Muir Wood, A. M. (2000) *Tunnelling: Management by Design*, Spon, London.

Muir Wood, A. M. and Duffy, F. (1991) Society's needs, in *Education for the Built Environment*, Seminar, Madingley Hall, Cambridge, Arup.

Muir Wood, A. M. and Fleming, C. A. (1981) *Coastal Hydraulics*, 2nd Edn, Macmillan, London.

Muir Wood, A. M. and Gibb, F. R. (1971) Design and construction of the Cargo Tunnel at Heathrow Airport, London, *Proc. Inst. Civ. Engrs*, **48**(2), 11–34.

Needham, J. (1971) *Scie. Civil. China*, (3) sec. 28. Cambridge University Press, Cambridge.

OECD (1971) *Conclusions (also Reports and Proceedings)*, Advisory Conference on Tunnelling, Organisation for Economic Cooperation and Development, Paris.

Olivier, H. J. (1970) Notes on the geological and hydrological investigations of the flooded Shaft 2, Orange-Fish Tunnel, Northern Cape, *Proc 13th Annual Congress, Geol. Soc., South Africa*.

O'Neill, O. S. (2002) *A Question of Trust*, BBC Reith Lectures.

Peck, R. B. (1969) Advantages and limitations of the Observational Method applied to soil mechanics, *Géotechnique*, **19**(2), 171–87.

Playfair, L. (1855) *The Study of Abstract Science Essential to the Progress of Industry* (Introductory Lecture to the Government School of Mines 1851) in *Literary Addresses*, Richard Griffin & Co., Glasgow, 47–86.

Poincaré, H. (1908) L'invention mathématique, in *Science et méthode*, 43–63 Flammarion, Paris.

Porter, R. (2000) *Enlightenment*, Allen Lane, London.

Powderham, A. J. (1994) An overview of the observational method: development in cut and cover tunnel projects, *Géotechnique*, **44**(4), 619–36.

Quinn, A. (1980) *Francis Bacon*, Oxford University Press, Oxford.

RAE (2000) *The Universe of Engineering – a UK Perspective*, Report of Working Group, Chairman, R. Malpas, Royal Academy of Engineering, London.

Reed, C. B. (1999) Choosing the right contract and the right contractor, *Proc. Øresund Link Dredging & Reclamation Conf.*, 73–86 Øresundskonsortiet, Copenhagen.

Reeve, D., Chadwick, A. and Fleming, C. (in press) *Coastal Engineering*, Spon, London.

Rennie, J. (1846). Presidential Address. *Min. Proc. Inst. Civ. Engrs*. **5**, 19–122.

Rennie, J. (1847). Presidential Address. *Min. Proc. Inst. Civ. Engrs*. **6**, 19–31.

RIBA (1992) *Report of the Steering Group on Architectural Education*, Royal Institute of British Architects, London.

Rimington, J. D. (1993) *Coping with technological risk: a 21st century problem*, The CSE Lecture, Royal Academy of Engineering, London.

Rothschild, N. M. U. (1971) *A Framework for Government Research and Development*, HMSO, London.

Rotter, J. M. (2001) *Guide for the Economic Design of Circular Metal Silos*, Spon, London.

Royal Society (1994) *Disposal of Radioactive Wastes in Deep Repositories*, Royal Society, London.

Segal Quince (1985) *The Cambridge Phenomenon*, Segal, Quince and Partners, Cambridge.

References

Skempton, A. W. (1959) Cast in-situ bored piles in London clay, *Géotechnique* **9**(4), 153–73.

Smiles, S. (1874) *Lives of the Engineers. Smeaton and Rennie*, John Murray, London.

SRC (1981) Long-term Research and Development Requirements in Civil Engineering, report prepared by Civil Engineering Task Force for the Sponsors, Science Research Council and Departments of Environment and Transport, CIRIA.

Stephenson, R. (1856) Presidential Address, *Min. Proc. Inst. Civ. Engrs*, **15**, 123–156.

Strategy Unit (2002) *Risk: Improving Government's Capability to Handle Risk and Uncertainty*, Summary Report, Strategy Unit, Cabinet Office, London.

Sutherland, R. and Pozzi, S. (1995) *The Changing Mathematical Background of Undergraduate Engineers: A Review of the Issues*, Engineering Council, London.

Tawney, R. H. (1921) *The Acquisitive Society*, Allen and Unwin, London.

Teich, A. H. and Runkle, D. C. (2000) The US experience with court-appointed scientific experts, *Science and Technology in the Eye of the Law*, 81–92, Royal Society, London.

Terris, A. K. and Morgan, H. D. (1961) New tunnels near Potters Bar in the Eastern Region of British Railways, *Proc. Inst. Civ. Engrs*, **18**, 289–304.

Thompson, E. P. (1970) *Warwick University Ltd*. Penguin Books, Harmondsworth.

Thring, M. W. (1980) *The Engineer's Conscience*, Northgate, London.

Toms, A. H. (1946) Folkestone Warren landslips: research carried out in 1939 by the Southern Railway Company, *Proc. Inst. Civ. Engrs Railway Paper No. 19*.

Trenter, N. A. and Warren, C. D. (1996) Further investigations at Folkestone Warren, *Géotechnique*, **46**(4), 589–620.

Unwin, W. C. (1911) Presidential address. *Proc. Inst. Civ. Engrs*, **187**, 2–27.

Viggiani, C. (2001) Does engineering need science?, in *Constitutive Modelling of Granular Materials*, by D. Kolymbas (ed.), 25–36, Springer, London.

Walker, E. and Simmons, R. E. (1847) *Report to the Commissioners of Railways on the Fatal Accident of 24th May 1847*, Commissioners of Railways, London.

Walker, J. (1841) Presidential address, *Proc. Inst. Civ. Engrs*, **1**, 15–23.

Watson, J. G. (1982, 1988) *The Institution of Civil Engineers: A Short History*, Thomas Telford, London.

Watt Committee (1999) *Watt Committee Energy Series Report 31: Energy Demand and Planning*, Spon, London.

Webb, H. A. and Ashwell, D. E. (1959) *Mathematical Tool-kit for Engineers*, Longmans, London.

Zienkiewicz, O. C. (1967, 71, 77, 87, 89 etc.) *The Finite Element Method* McGraw-Hill, New York.

Index

Note: page numbers in *italics* indicate Figures, 'AMW' = 'Sir Alan Muir Wood'

accident investigations
   dissemination of findings, 82
   political overreaction, 189
   resulting sanctions, 146
adjudicator
   Latham's recommendations, 33
   role of Engineer in ICE Conditions, 28, 61
Advisory Committee on Mathematics, 160
Advisory Committee on Research and Development (ACORD, Department of Energy), 195–196
Advisory Council on Applied Research and Development (ACARD, Cabinet Office), 196–197
affordable housing, 217
Age of Enlightenment, 164
Alkali Act (1863), 210, 211
American Society of Civil Engineers (ASCE), publications, 54
Appropriate Development Forum, 152
Archimedes, 8
architect(s)
   compared with engineers, 86, 87, 90–91, 149–150
   example of capability/knowledge levels, *91*
   RIBA Steering Group on education, 85–87

armed services
   works for, 146
   see also Royal Navy
Ashby, Lord (Eric), Graham Clark Lecture (1978), 210, 211, 220
attribution of success, 48
awareness (capability/knowledge level), 58, 90, 110
   examples for architects and engineers, *91*

background of early engineers, 17
Bacon, Francis, 13
Bacon, Roger, 13
Bagot, Robin, 194, 195
Baker's Bell, 21
Banwell Report, 29
Barton-on-Sea, coastal protection system, 101–102, 136
Bazalgette, Sir Joseph, 24
beach sediments
   modelling of movement, 107–108
   observational approach, 98–99
best practicable means (bpm), 210
   and local controls, 211, 212
bitumen, early use of, 7
Bondi, Sir Hermann, 133
'bouncing bomb', 143
Brassey, Thomas, 23
bridges
   collapses, 110, 146

Civil engineering in context

bridges (*continued*)
  Inglis Bridge, 116
  models, 104
  railway, 123
Brindley, James, 11, 23
Brinkworth, Professor B. J., 208
Britannia Bridge (Menai Straits), 104
British Constructional Steelwork Association (BCSA), guidance on design of bunkers and silos, 142
British Materials Handling Association, silo design guides, 142
Brunel, Sir Marc Isambard, 14, 23, 24, 104, 138, 139
Brunelleschi, Filippo, 11
build–operate–transfer (BOT) projects, early examples, 23
Building Research Station (BRS), history of development, 147–148
Bullock, Christopher, 197
By, John, 4
Byzantine domes, 11

Cabinet Office
  Advisory Council on Applied Research and Development, 196–197
  Strategy Unit, 201
  report on risk control, 91–92
Cairncross, Sir Alec, report on 1970–75 Channel Tunnel project, 170
Cambridge University, 196
  AMW at, 115–116
capability/knowledge levels
  comparison of related professions, 90–91, *91*
  for various project-team members, 57–58
Cartesian philosophy, 13, 164
central command economy, 202
Central Electricity Generating Board (CEGB)
  design of coal bunkers, 142
  siting of Dungeness Power Station, 137–138
Centre for Construction Law and Management (CCLM), 60
Channel Expressway, 171, 172

Channel Tunnel, 166–181
  1958–60 studies, 166–167
  1964–65 studies, 167–169
  1970–75 studies, 170
  1987–98 construction phase, 172–181
  alternative schemes (1985 competition), 171–172
  Beaumont trial tunnel, 179
  cost and time overruns, 175, 194
  differences in construction organisation, 181
  Disputes Panel, 76, 172–174
  early history, 178–179
  factory-made sections, 35
  financing of, 171, 175
  geological surveys, 166–167, 168–169, 178
  Government's lack of understanding, 175
  opposition to, 179
  Øresund Link project compared with, 76, 177–178
  procurement procedural defects, 175–177, 181
  technical aspects, 179–181
Channel Tunnel Study Group (CTSG), 166, 167
Chartered Engineer (CEng), 46, 50
  qualifications, 159
Chatham Naval Dockyard, 114, 146
chemical engineering, 18–19
civil engineer
  changes in scope of work, 186
  code of professional conduct, 155–158, 187–188, 191
  definition, 150
  example of capability/knowledge levels, *91*
  repositioning of, 150–152
  see also engineer(s)
civil engineering
  AMW's first job, 121–123
  AMW's reason for choosing as career, 113
  France compared with Britain, 13–14, 164–165
  as observational discipline, 18

228

satisfaction as career, 152–155
Civil Engineering Research Council/
    Association, 148
Civil Procedure Rules (CPR), 77
'claims engineers', on Channel Tunnel
    project, 174
Clean Air Act (1956), 210–211
Clyde Tunnel, 97, 129–130
coal bunker collapse, 140–142
coastal power station, design of cooling
    water works, 108, 136–137
coastal works
    AMW's consulting experience,
        98–99, 134–138
    'soft' protection approach, 136
Cockcroft Report, 159
code of conduct, engineer's, 155–158
collaborative working, 35, 66, 94
colleges, role in unified engineering
    profession, 50–51, 151
Colorado River (USA), hydro-power
    plants, 210
commercial approach, 30–31
    defects of, 64, 71, 192
    dispute resolution in, 75–76
Commission de Surveillance (in Channel
    Tunnel 1964–65 investigations),
    167, 168
communications, project, 72
communicator
    architect as, 150
    engineer as, 150–151, 192–193
competence (capability/knowledge
    level), 58, 90, 110
    boundaries recognised by engineer,
        156, 158
    examples for architects and engineers,
        91
Computational Fluid Dynamics (CFD),
    108
conference proceedings, 56
Construction Design Management
    (CDM), 33
Construction Industry Research and
    Information Association (CIRIA),
    148
    Underwater Engineering Group,
        148

construction management contracts,
    sharing of risk between Client and
    Contractor, 65
consultants' boards, 133–134, 154
consultants register, Latham's
    recommendations, 33
consulting engineers, 49, 61
continuity in projects, 31, 92
    disruption being unhelpful, 31
    provisions in red FIDIC, 38
    ways of achieving, 31, 36
Contract, objectives, 155
Contract Conditions
    getting best out of, 66–73
    see also FIDIC Contract Documents;
        ICE Conditions of Contract;
        New Engineering Contract
Contractor
    loading of uncertainty cost onto, 64
    relationship with Engineer (in ICE
        Conditions), 24, 25, 28, 62
    sharing of risk with Client, 64–66
convergent thinking, 90
Conway Bridge, 104
corporate social responsibility (CSR),
    184
cost-reimbursable contracts, sharing of
    risk between Client and
    Contractor, 65
Council of Engineering Institutions
    (CEI), 45
Court Appointed Scientific Experts
    (CASE) witnesses, 81
Ctesibus, 8
Cuilfail road tunnel (Lewes), 132
customer-based approach to projects, 35,
    85, 203

Dee Bridge, collapse, 146
Deep Earth Club, proposal for storage of
    nuclear waste, 132–133
'descriptive engineer', 57
design
    as centre of professional education
        and training, 85–87
    for future, 109–110
    meaning of term, 85, 86
    systems approach, 88–89

229

Civil engineering in context

design (*continued*)
  trial approach to, 88
design–build–finance–transfer (DBFI) schemes, 207
Design Checker, 110–112
design-and-build projects
  early examples, 23
  sharing of risk between Client and Contractor, 65
design-and-manage projects, sharing of risk between Client and Contractor, 65
designer
  Engineer (in ICE Conditions) as, 24, 28, 62, 70
  in Highways Agency ECI Scheme, 42
  Latham's recommendations, 32
  role in NEC, 36–37
developing countries, funds for, 193
developing countries, professionals from, 216
development projects, degradation due to, 189
direct-labour projects, 23
disabled people access legislation, effects, 212
dispute resolution
  Engineer's approach, 28, 76
  external arrangements, 75
  legal approach, 75–76
dispute-resolution methods, 33–34
  Channel Tunnel, 76, 172–174
  Øresund Link project, 76, 177–178
divergent thinking, 90
Dominican Republic, beach erosion, 98
Dreyfus, Gilbert, 172
Duffy, Dr Francis, 83, 85
Dungeness A Nuclear Power Station, 136–138

Early Contractor Involvement (ECI) Scheme, 42, 94
earth tides, 130–132
  Orange-Fish Tunnel data, 130, *131*
Economic and Social Research Council (ESRC), joint SERC/ESRC liaison committee, 200

economy
  factors affecting, 214–216
  middle way, 202, 214
education and training of engineers, 14–15, 17–18, 50–51, 158–163
  Britain compared with France, 14
educational reasons for studying history, 3–4
efficient engineering, and ethics, 187
Egan Report, 34–35, 85, 203
  Highways Agency ECI Scheme based on, 42
Egyptians, 7
electrical engineering, 18, 20–21
electrical industry, reasons for decline, 214
electronic engineering, 20, 21
energy, 208–210
  engineers' advice on, 208
energy ratio, variation during industrial development, 208, 209
energy strategy, 209–210
  factors affecting, 186–187
engineer(s)
  characteristics, 190, 192
  code of conduct, 155–158
  compared with architects, 86, 87, 90–91, 149–150
  compared with politicians, 190
  education compared between Britain and France, 14
  example of capability/knowledge levels, 91
  factors affecting shortage of, 46–47
  own recognition of boundary for competence, 156, 158
  public credit lacking for success, 48
  role and functions, 89–90, 109, 145
  statutory registration of, 51, 157
Engineer (in Highways Agency ECI Scheme), 42
Engineer (in ICE Conditions of Contract)
  appointment of, 66–67
  as arbiter/interpreter, 24, 28
  as designer, 24, 28, 62, 70
  incompetent, 26, 28

Index

potential conflict of duties, 28, 35–36, 61
replacement in NEC, 35–36
restrictions on independence, 28, 62
role and functions, 24–25, 28, 35–36, 70–71
as supervisor, 24, 28, 70
terms of reference, 26, 67
Engineer (in red FIDIC)
continuity provided by, 38
independence, 38, 39
role and functions, 38–39
engineering
'banausic' (physical work) stigma, 8, 17
common ground between various disciplines, 51, 109–110
disciplines within, 18–21
widening scope, 43–49
Engineering and Construction Contract (ECC) *see* New Engineering Contract (NEC)
Engineering Council (EC), 45
on entry standards for engineering courses, 46
on mathematics, 159–160
engineering departments, first established in Britain, 17–18
engineering ethics, 184–189
as covenant, 188
engineering institutions, 15, 43–59
as CEI/EC members, 45
rivalry between, 44
see also Institution of Civil Engineers; Institution of Mechanical Engineers; Institution of Structural Engineers
engineering profession
advice to Government, 202
early developments, 15–18
education of public as to role, 212–213
on energy, 208
Finniston Commission on, 44–45, 51
historical perspective, 58
isolation of, 148–149
qualifications for membership, 57–58
statutory regulation of, 51, 157

unification of, 49–51, 203
AMW's proposal, 50–51, 151
Engineering and Technology Board (ETB), 45–46, 212
'enlightened purchaser', 154, 203
see also Intelligent Market concept
environmental considerations, 109, 189, 210–212
environmental impact assessment (EIA), 189, 210
environmental legislation, 210–212
applicability, 211
environometrics, 185
escrow, 72
Essig, Philippe, 174
ethical investments, 184
ethics, 183–184
engineering, 184–189
meaning of term, 183
not covered in multi-consultant projects, 191
professional vs personal, 184–185
relationship to politics, 185, 190–194
teaching in schools, 217
Eurobridge, 171
Euroroute 'hybrid' scheme, 171–172
Eurotunnel, 172
Eurotunnel (ET) company, 173
Chief Executive, 174
Evans, J. T., 126
expert witness(es), 77–82
area of competence, 77, 82
cost saving by, 81
failure to present fair view, 78
language used by, 77
legal procedures, 79
and other witnesses, 79–81
rapporteur to represent, 81
roles and functions, 74–75, 79, 82
'without prejudice' meetings, 80
expertise (capability/knowledge level), 58, 90, 110
examples for architects and engineers, 91

factory-made construction components, 35

231

Fairclough, Dr J., 50
FIDIC Contract Documents, 37–41
   Dispute Adjudication Board, 39
   Red document (on construction projects), 37–41
     and criteria for success, 41
     criticisms, 41
     provisions for continuity, 38
     provisions for uncertainty, 40–41
     relationships between Parties, 28–29
     on value engineering, 39–40
   role of Engineer, 38–39
   Silver document (on EPC/turnkey projects), 37
Fiji, hydro-electric project, 97–98
finance
   for Channel Tunnel, 175
   factors affecting, 173, 193
   private, 22, 204
   see also PFI projects
financial sector, compared with industrial sector, 213
Finniston Commission, 44–45, 51
Finniston Report, 45, 157
first engineering appointment to project, 154
Fleming, Dr C. A., 108
flint tools, 7
Folkestone Leas Cliff, remedial works, 134–135
Folkestone Warren coastal landslides, 124–125, 135
   borehole in Channel Tunnel investigations, 166–167
   early history, 125
   movement/slip indicators, 135
Forensic Practitioners, Council for Registration of, 77
fossil markers, 124
France
   approach to civil engineering compared with that in Britain, 164–165
   early civil engineering compared with that in Britain, 13–14
   law, 174
free market economy, 202, 214, 215

French–British collaboration
   on Channel Tunnel, 166–181
   revival in 1970, 170
future, dependency on past, 5

Gateshead Millennium Bridge, 48
Gautheron, Jean, 172
GC/Works 1 form of contract, 28
General Electric Company (GEC), 214
geodetic construction, 143
Geological Survey, 124, 133
Geotechnical Baseline Report (GBR), 68
   see also Reference Conditions
geotechnical models, 104, 107
Géotechnique (journal), 56–57
'golden age' of engineering, 12–21
good practice, application of principles, 63–66
Government strategies
   on energy, 186–187
   non-implementation of, 92
Governmental ineptitude, on Channel Tunnel project, 175
Governmental research funding, 200–201
Greece, marina project, 98–99, 136
Greeks, 7–9
   attitude to practical skills, 8
   fascination with theory, 7
'green' movement, 149
   attitude to engineers, 149
gross domestic product (GDP), factors affecting, 215–216, 218
ground investigations, interpretation of, 67–68
ground models, 104

hagiographic reasons for studying history, 1–2
Halcrow, Sir William, 124
Halcrow, Sir William, & Partners, AMW joining, 127
Hambly, E. C., 53
Harding, Sir Harold, 52, 135, 166, 169
hazard, meaning of term, 63
Heathrow *see* London Heathrow...
Henley Regatta, demountable suspension footbridge, 144, 172

Index

'heritage sites', valuation of, 185
Hero/Heron, 8
Higher National Diploma (HND), 47
Highways Agency, Early Contractor Involvement (ECI) Scheme, 42, 94
historical buildings, valuation of, 185–186
history
   reasons for studying, 1–4
   uses, 4–6
HMS Newcastle, 121
HMS Paladin, 120
HMS Petard, 118, 119–120
holistic approach of Proof Engineer, 112
holistic design process, 87–89, 191
Honduras, coastal lake, 98
House of Lords, 192
hydraulic models, 104
hydraulics research, AMW's early activities, 126–127
Hydraulics Research Station, 148
hydro-power, environmental impact of, 188, 210
hydrogen, as fuel for transport, 196

ICE *see* Institution of Civil Engineers
ICE Conditions of Contract
   appointment of Engineer, 66–67
   criteria for success, 27
   criticisms, 27–29
   decline of, 62–63
   first published, 24
   as model contract, 29
   Parties involved, 27, 72
   reasons for failure, 25–26, 27–29
   requirements for Engineer, 27–28
   restrictions on independence of Engineer, 28
   role and functions of Engineer, 24–26, 28
   successful projects using, 61, 70
immigration control, 216
Incorporated Engineer (IEng), 46, 50
   qualifications, 159
Industrial Revolution, 10–12
industry
   compared with profession, 182
   and higher education, 196

infrastructure, need for long-term policies, 193, 207
Inglis, Sir Charles, 37, 116, 158
Inglis, Sir Claude, 148
inland waterways, AMW's research, 126–127
innovation, factors affecting, 164
Institution of Civil Engineers (ICE)
   on acceptance of degree course, 163
   Annual Review 2000, 5, 53
   on background of early members, 17
   change in by-laws, 59
   Director-General, 59
   engineers' names in ICE building, 1–2
   evidence to Finniston Commission, 44–45
   evolution of, 51–53
   functions, 59
   no help given to membership after 1939–45/46 war, 121
   Presidential Address 1845/46 (Rennie), 12
   Presidential Address 1846/47 (Rennie), 12, 14
   Presidential Address 1855/56 (Stephenson), 15
   Presidential Address 1910/11 (Unwin), 14
   Presidential Address 1940/41 (Inglis), 116
   Presidential Address 1977/78 (Muir Wood), 7, 53, 58, 203
   President's role, 59
   publications/reports policy, 54–57
   qualifications for membership, 57–58, 158
   Rules of Professional Conduct, 156, 187–188, 191
   special interest groups, 52–53
   and technological change (in 1970s), 52
   widening of range of acceptable disciplines and occupations, 191
Institution of Mechanical Engineers, 15, 17, 152
Institution of Structural Engineers, 116, 159
   see also Joint Board of Moderators

233

institutions, 15, 43–59
Intelligent Market concept, 26–27, 84, 154–155, 202–203
  see also partnering
interactive numerical models, 107–108
Intermediate Technology Development Group (ITDG), 151–152
international associations, 147, 197–199
International Baccalaureate, 159
International Conference on Soil Mechanics and Foundation Engineering (New Delhi, 1994), 125
International Tunnelling Association (ITA), 197–199
  Honorary Life President, 199
  inaugural meeting, 198
  journal, 198
  membership, 199
  origins, 197
  Silver Jubilee, 199
  Working Groups, 198–199
Istanbul, St Sophia church, 11

Joint Board of Moderators (for ICE and IStructE), 116, 159, 163
Jubilee Line Extension, 64

landslides
  AMW's experience, 124–125, 135
  Folkestone Warren, 124–125, 135
  Potters Bar railway tunnel spoil, 96–97
Latham Report, 29–34
  criticisms, 30, 31, 32
  terms of reference, 30
  'traditional method' portrayed in, 32
law
  France compared with England, 174
  pre-project considerations, 73–75
  in project disputes, 75–77
law of unintended consequences, 201, 210
lawyers
  'complaints' about NEC, 35
  privileges and pay, 182–183
  role of, 60, 219
lead designer, as Project Manager, 33

learned society journals, 55
legislation
  danger of rigid standards in, 212, 220
  professional judgement to be allowed under, 211, 220
Levens Park, 194–195
liquidity index, calculation for London Clay spoil, 96
litigation, effects of, 3, 6
Little, Michael, 172
Locke, Joseph, 23
London Heathrow Airport
  Cargo Tunnel, 129
  rail access proposals, 143–144
  Terminal 5, 35, 73
London Heathrow Express, tunnel collapse, 101, 187
London Underground, Victoria Line tunnels, 128–129
Lowestoft, sea wall, 135–136
loyalty, 49, 153
lump sum contract
  for Channel Tunnel terminals, 173
  sharing of risk between Client and Contractor, 65

M6 motorway, Kendal spur, 194–195
Malcor, Réné, 166
Malinvaud, Philippe, 172
management contracts, sharing of risk between Client and Contractor, 65
Manzoni, Sir Herbert, 148
marine engineering, AMW's experience, 117–121
Maslow's pyramid of satisfactions, 215
mathematics
  required when using computer methods, 106, 158
  requirements for engineers, 160–161
  teaching in schools, 159, 160
  university courses involving, 115–116, 158–159, 163
mechanical engineering, 19–20
  relationship with civil engineering, 15–16
media, attitude to engineers, 48
mediaeval engineering, 11

Index

Mediterranean, AMW during War, 117–119
Menai Straits
  Stephenson's tubular bridge, 104
  Telford's road bridge, 12
metals, early technology, 7
Metropolitan Board of Works (London), standard form of contract, 24
Mill, John Stuart, 215
millwright, as forerunner of civil engineer, 11
models, 103–109
  accuracy, 105
  examples, 104
  validation of, 106
  verification of, 105–106
morals, meaning of term, 183
Morgan, Horace D., 124, 127–128, 143
motivation of engineers, 116, 162
motoring, factors affecting safety, 187
motorway planning inquiries, 194–195
Muir Wood, Sir Alan
  on ACARD (Cabinet Office), 196–197
  on ACORD (Department of Energy), 195–196
  as advisor to Parliamentary Transport Committee, 171
  on appropriate technology and 'green' issues, 151–152
  at Folkestone Warren, 124–125
  beach sediment movement model, 107–108
  Channel Tunnel experience, 166–181
  consulting experience, 127–144
    bridges, 144
    coastal and riverine works, 98–99, 134–139
    landslides, 135
    railways, 143–144
    silos, 139–142
    tunnelling, 96–98, 127–134, 166–181
  on Design Checker/Proof Engineer, 111
  early life, 114–115
  Education for the Built Environment seminar, 83
  Greek marina project, 98–99, 136
  ICE Presidential Address (1977/78), 7, 53, 58, 203
  inland waterways research, 126–127
  and International Tunnelling Association, 197–199
  joined Halcrow, 127
  in Navy, 117–121
  political experiences, 194–201
  reason for choosing civil engineering as career, 113
  on research councils, 199–201
  on RIBA Steering Group on Architectural Education, 85–87
  with Southern Railway Company, 121–125
  university course, 115–116
  Unwin Lecture (1982), 208
  wife, 121
multi-consultant projects, ethical issues not covered in, 191

National Audit Office, 203
New Crane Wharf (Wapping, London), 139
New Engineering Contract (NEC), 33, 35–37
  compensation events, 36
  criticisms, 35–36
  designer's role and functions, 36
  legal language in, 35, 66
Newcomen, Thomas, 19
Nuclear Industry Radioactive Waste Executive (NIREX), 133
nuclear power, as part of energy policy, 186
nuclear warfare, 186
nuclear waste, disposal/storage of, 132–133, 186
numerical models, 105–109
  engineer's input, 106, 108–109
  interactive, 107–108

observation, 95–96
Observational Method (OM), 96, 99–103
  early use, 99
  examples of application, 96–99, 101–102

Observation Method (OM) (*continued*)
  features affecting successful application, 103
  progressive modification approach, 99
  reasons for failure, 103
OECD, Advisory Conference on Tunnelling, 197
offshore engineering, 44
offshore wind and wave energy, ACORD discussions, 195–196
O'Neill, Onora, on trust, 61
optimisation, Channel Tunnel project, 175–176
Orange-Fish Tunnel, 99–100, 130, *131*
Øresund Link project
  Disputes Advisory Boards, 76, 177–178
  factory-made elements, 35
  Reference Conditions, 68, 178
Orski, C. K., 197
overcrowded areas (of UK), moving away from, 217

Pantheon (Rome), dome, 10
Parliamentary Bills, 15, 74, 129
Parliamentary Transport Committee, AMW as advisor, 171
Parties, relationships between
  effect of risk aversion, 64
  in ICE Conditions, 25, 28, 62
  in red FIDIC, 28–29
partnering, 33, 34, 40, 42, 73, 155, 204
  sharing of risk between Client and Contractor, 65
  see also Intelligent Market concept
past, effect on future, 5
pay of engineers compared with other professions, 46–47
Peck, Ralph B., 99
Permanent International Association of Road Congresses (PIARC), 147
  Technical Committee for Tunnelling, 197
personal ambition, 48, 153
  factors affecting, 48–49
personal debt, 217–218
PFI projects, 204–206
  excessive costs, 105, 194
  factors affecting costs, 205, 206
  objectives, 205
  'off-account' value (in 2003), 205
  railway infrastructure projects, 207
  risk transfer in, 205–206
pioneering engineers, 3
  educational standards, 14, 163
  still needed, 6
Pisa, leaning tower, protection of, 48
planning processes, 74
Plato, 8
Playfair, Lyon, 16, 20
Poincaré, Henri, 90
politicians
  attitudes to engineers, 4, 192
  characteristics, 190, 192, 213
  compared with engineers, 190
  reluctance to share privileges, 157
Ponts et Chaussées organisation (France), 13, 15, 22
Potters Bar railway tunnels, 96–97, 128
practice
  relationship to theory, 4
  Romans' concentration on, 7, 9
pragmatism, 163–165
  as barrier to innovation, 164
presentation, no substitute for technical abilities, 192
prison population, 214
Private Finance Initiative *see* PFI projects
private funding for infrastructure projects, 22, 204
privatisation, 204
process–product interrelationships, 94–95
procurement methods
  for Channel Tunnel, 173, 175
  early methods, 22–24
  'traditional method', 24–27
profession, features, 182
professional engineer, qualifications, 57–58, 158
professional relationships, 49
professionalism, role in 'traditional method', 27
Professionals for the Built Environment (PBEs), functions, 83–84

Index

project cost estimates, factors affecting, 93
project definition, 92–95
  Channel Tunnel project, 173–174
project fragmentation
  in Channel Tunnel project, 173
  effects, 191, 192
Project Manager
  lead designer acting as, 33
  in NEC ECC, 35, 36
project report, 72
Proof Engineer, 110–112
  factors affecting increased dependence on, 111
  holistic approach, 112
  terms of reference, 112
property values, inflated, 217
public interest, 201–210
publications, ICE's policy on, 54–57
purposeful continuity in projects, 31, 92

quality assurance (QA), 95
  Latham on, 33
  red FIDIC on, 40

radioactive wastes, disposal of, 132–133
rails, temperature effects on long sections, 122–123
railway bridges, AMW surveying, 123
railway canteens, 123
Railway Law, Stephenson's complaints, 15, 207
railway maintenance, 121–123
railway management, Robert Stephenson on, 15, 143
railways
  mismanagement of, 206–207
  nationalisation of, 206
  privatisation of, 207
  regulatory organisations, 207
  threat of renationalisation, 207
Rama, Marcel, 166
Rankine active state loading, coal bunker design based on, 140, 142
records
  of decision-making processes, 165
  Thames Tunnel construction, 139
refereeing of papers, 56, 57

reference books, 56
Reference Conditions, 68
  use in Øresund Link project, 68, 178
regional government, advantages, 193
Register of Engineers for Disaster Relief (REDR), 152
registered engineers, 157–158
renewable energy sources, cost comparison, 208
Rennie, John (1761–1821), 11, 23
Rennie, Sir John (1794–1874), 12, 14
reports about construction, 29–35
research councils, 199–201
research and development, expenditure for construction sector, 34, 200
Resident Engineer, 23
RIBA Steering Group on Architectural Education, 85–87
Rideau Canal (Canada), 4
Ridley, Tony M., 53, 174
Rimington, John, 189
risk
  meaning of term, 63, 151
  sharing between Client and Contractor, 64–66
risk aversion
  costs, 63
  effects, 220
risk control pyramid, 92
risk management, 154
  Channel Tunnel project, 175
  and human behaviour, 187
  as part of engineer's culture, 156–157
risk procedures, 71, 91–92
risk transfer, in PFI projects, 205–206
road construction research, 147
Robertson, V. A. M., 124, 127
rock tunnelling, steel arch tunnel supports used, 132
Romans
  practical skills, 7, 9–10
  water supply systems, 9, 10
Royal Academy of Engineering (RAE), 45, 46, 204
Royal Engineers, 144
Royal Military College, PFI project, 206
Royal Navy, AMW in, 117–121, 146

237

Civil engineering in context

Royal Society, 13
   on A-level mathematics curriculum, 159
   working party on disposal of radioactive wastes, 133

Sacks, Jonathan, 188
safety, negative effects of excessively expensive protection, 189
Sainsbury, R. N., 53
São Paulo, interceptor tunnel, 133–134
schedule-of-rates contracts, sharing of risk between Client and Contractor, 65
schools, teaching of mathematics, 159, 160
Schumacher, E. F., 151
Science and Engineering Research Council (SERC), 199–200
Science Research Council (SRC), report on long-term R&D requirements in civil engineering, 200–201
scientific evidence, 82
Scottish military roads, 14
Second Severn Crossing, 35
'seed-corn' fund, 196
services engineer, example of capability/knowledge levels, 91
Sharlston Colliery, collapse of coal bunker, 140–142
ship's damage control diagram, 119, 120
shipyard practices, 117, 146
Sierra Club environmental organisation, 210
'signing-off', disadvantages of practice, 32–33, 103
silos
   AMW's consulting experience, 139–142
   engineering disciplines involved in design, 139
site investigation, 67–68
   records, 69
slip indicators, Folkestone Warren, 135
Smeaton, John, 23
Smiles, Samuel, on background of engineers, 17
smoke control zones, 211

social cohesion, factors affecting, 213–214
social considerations, 109, 189
social structure, 202
'soft' coastal protection, 136
Southern Railway Company, 121–123
Squires, R. H., 142
stakeholders, relationships between, 60–82, 319
standard forms of contract
   Latham's survey on attitudes to, 33
   see also FIDIC Contract Documents; ICE Conditions of Contract; New Engineering Contract
state-commissioned work, 13–14, 22
statutory regulation of engineering profession, 51, 157
steel arch tunnel supports, 132
Stephenson, George, 17
Stephenson, Robert, 15, 104
   Dee Bridge designed by, 146
   on railway management, 15, 143
stone arches, early development, 10
Stradling, Reginald, 148
strategic decisions, factors affecting implementation of, 92
structural models, 104, 107
subsistence economy, 215
Surrogate Operator, 31, 41
sustainability, 212
   collision with market forces, 215
   engineer's responsibilities, 58, 203
Sydney Water Board, Ocean Outfall project, 134, 154
systems engineering, 83–92
   absence in design of silos, 139
   at heart of engineer's work, 161

Tagore, Rabindranath, 217
Target Contract, 42, 73
   in Channel Tunnel project, 173
Tawney, R. H., 182
teamwork
   change in US industry, 34
   emphasis in Egan Report, 34
   emphasis in NEC, 36
technological change, effects (1970s), 52
telephone exchange tunnels, 127–128

Index

Telford, Thomas, 12, 23
tender procedures, 68–69
Thames (Rotherhithe–Wapping) Tunnel, 23, 24, 104
　Duke of Wellington's support for, 179
　protection works, 138–139
Thatcher, Margaret, 144
theory
　Greeks' fascination with, 7
　relationship to practice, 4
Thessalonica (Greece), marina project, 98–99, 136
Thomé de Gamond, 178
Thompson, E. P., 196
Thornycroft's shipyard (Southampton), AMW at, 117
three-dimensional models, in tunnelling, 107
Thring, Professor Meredith, 152
timber baulks, off-loading from ships, 114–115
Toms, A. H., 122, 124
tool-kits, mathematical, 162
'traditional method' of project procurement and management, 24–27
　criteria for success, 25, 27
　portrayal in Latham Report, 32
　reasons for failure, 25–26, 27
　role and functions of Engineer, 24–26
　see also ICE Conditions of Contract
training of engineers, 14–15, 17–18, 50–51
Trans-Manche Link (TML) contractors, 173
　Chairman, 174
Transport and Road Research Laboratory (TRRL), history of development, 147
Tredgold, Thomas, 150
triggers, in Observational Method, 100–101
trust, 61
　Engineer's role, 25, 28, 62–63, 212–213, 220
Trusts, works financed by, 22, 204
tunnel boring machines (TBMs), in Channel Tunnel, 180

tunnel convergence
　in Channel Tunnel, 180
　monitoring of, 99–100, 100–101
tunnelling
　AMW's consulting experience, 96–98, 127–134, 166–181
　Design Checker/Proof Engineer, 111
　three-dimensional models, 107
　see also British Tunnelling Society; Channel Tunnel; International Tunnelling Association
turnkey projects
　FIDIC Contract Documents, 37
　sharing of risk between Client and Contractor, 65
twenty-first century engineer, 145–165
　historical legacy, 145–150
　implications for education and training, 158–163

uncertainty
　loading of cost onto Contractor, 64
　meaning of term, 63
underground planning, 199
unintended consequences, law of, 201, 210
union demarcation, in shipyards, 117
university
　AMW at, 115–116
　engineering departments first established, 17–18
Unwin, W. Cawthorne, 14
Unwin lecture (1982), 208
US construction industry, 34
US Navy ships, 120
utilitarian reasons for studying history, 2–3

value engineering, red FIDIC on, 39–40
variation(s), PFI project costs affected by, 206
Vickers factory (Weybridge), AMW at, 116, 143

Wallis, Dr Barnes, 143
Walton, E. W. K., 119
Warwick University, 196

239

water supply systems
    Greek (Pergamon), 8–9
    Roman, 9, 10
Watt Committee, on energy demand and planning, 208–209
Watt, James, 19–20
Webb, H. A., 115–116
Weinstock, Lord (Arnold), 214
Wellington, Duke of, 179
Wellington bombers, 116, 143
'whistle-blower's charter', 189
wind power, as part of energy policy, 195
'without prejudice' meetings (for expert witnesses), 80
    preliminary meeting for, 81
    timing and structure, 80
working hours, excessive, 48–49, 154, 183